h-Principles and Flexibility in Geometry

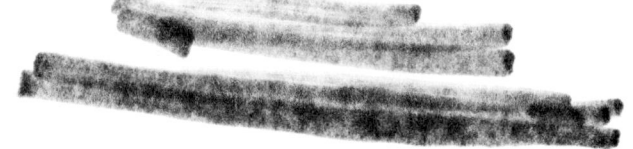

of the
American Mathematical Society

Number 779

h-Principles and Flexibility in Geometry

Hansjörg Geiges

July 2003 • Volume 164 • Number 779 (first of 5 numbers) • ISSN 0065-9266

American Mathematical Society
Providence, Rhode Island

2000 Mathematics Subject Classification. Primary 53C23; Secondary 57R42, 53Dxx.

Library of Congress Cataloging-in-Publication Data

Geiges, Hansjörg, 1966–
    h-principles and flexibility in geometry / Hansjörg Geiges.
        p. cm. — (Memoirs of the American Mathematical Society, ISSN 0065-9266 ; no. 779)
    "Volume 164, number 779 (first of 5 numbers)."
    Includes bibliographical references.
    ISBN 0-8218-3315-4 (alk. paper)
    1. Global differential geometry. 2. Immersions (Mathematics) 3. Symplectic manifolds.
I. Title. II. Series.

QA3.A57 no. 779
[QA670]
510 s–dc21
[516.3′62]   2003048025

Memoirs of the American Mathematical Society

This journal is devoted entirely to research in pure and applied mathematics.

Subscription information. The 2003 subscription begins with volume 161 and consists of six mailings, each containing one or more numbers. Subscription prices for 2003 are $555 list, $444 institutional member. A late charge of 10% of the subscription price will be imposed on orders received from nonmembers after January 1 of the subscription year. Subscribers outside the United States and India must pay a postage surcharge of $31; subscribers in India must pay a postage surcharge of $43. Expedited delivery to destinations in North America $35; elsewhere $130. Each number may be ordered separately; *please specify number* when ordering an individual number. For prices and titles of recently released numbers, see the New Publications sections of the *Notices of the American Mathematical Society*.

Back number information. For back issues see the *AMS Catalog of Publications*.

Subscriptions and orders should be addressed to the American Mathematical Society, P. O. Box 845904, Boston, MA 02284-5904, USA. *All orders must be accompanied by payment.* Other correspondence should be addressed to 201 Charles Street, Providence, RI 02904-2294, USA.

Copying and reprinting. Individual readers of this publication, and nonprofit libraries acting for them, are permitted to make fair use of the material, such as to copy a chapter for use in teaching or research. Permission is granted to quote brief passages from this publication in reviews, provided the customary acknowledgment of the source is given.

Republication, systematic copying, or multiple reproduction of any material in this publication is permitted only under license from the American Mathematical Society. Requests for such permission should be addressed to the Acquisitions Department, American Mathematical Society, 201 Charles Street, Providence, Rhode Island 02904-2294, USA. Requests can also be made by e-mail to reprint-permission@ams.org.

Memoirs of the American Mathematical Society is published bimonthly (each volume consisting usually of more than one number) by the American Mathematical Society at 201 Charles Street, Providence, RI 02904-2294, USA. Periodicals postage paid at Providence, RI. Postmaster: Send address changes to Memoirs, American Mathematical Society, 201 Charles Street, Providence, RI 02904-2294, USA.

© 2003 by the American Mathematical Society. All rights reserved.
This publication is indexed in *Science Citation Index*®, *SciSearch*®, *Research Alert*®, *CompuMath Citation Index*®, *Current Contents*®/*Physical, Chemical & Earth Sciences*.
Printed in the United States of America.

∞ The paper used in this book is acid-free and falls within the guidelines established to ensure permanence and durability.
Visit the AMS home page at http://www.ams.org/

10 9 8 7 6 5 4 3 2 1   08 07 06 05 04 03

Contents

Chapter 1. Introduction ... 1

Chapter 2. Differential Relations and h-Principles ... 3

Chapter 3. The h-Principle for open, invariant Relations ... 11
   3.1. Open, invariant relations ... 11
   3.2. Statement of the theorem ... 13
   3.3. Applications ... 14
   3.4. Proof of the theorem ... 24
   3.5. Further details of the proof ... 32

Chapter 4. Convex Integration Theory ... 43
   4.1. The h-principle for open, ample relations ... 44
   4.2. Proof of the simplest case ... 49
   4.3. Applications to symplectic and contact geometry ... 51

Bibliography ... 57

Abstract

The notion of homotopy principle or h-principle is one of the key concepts in an elegant language developed by Gromov to deal with a host of questions in geometry and topology. Roughly speaking, for a certain differential geometric problem to satisfy the h-principle is equivalent to saying that a solution to the problem exists whenever certain obvious topological obstructions vanish.

The foundational examples for applications of Gromov's ideas include

(i) Hirsch-Smale immersion theory,
(ii) Nash-Kuiper C^1–isometric immersion theory,
(iii) existence of symplectic and contact structures on open manifolds.

Gromov has developed several powerful methods that allow one to prove h-principles. These notes, based on lectures given in the Graduiertenkolleg of Leipzig University, present two such methods which are strong enough to deal with applications (i) and (iii).

Received by the editor October 2001.
2000 *Mathematics Subject Classification.* 53C23; 57R42, 53Dxx.

CHAPTER 1

Introduction

Here are two sample theorems that we are going to prove in these lectures. To state the first one, we introduce some standard notation.

Given manifolds M and N, we write $C^r(M, N)$ for the set of r times continuously differentiable maps from M to N (resp. continuous maps if $r = 0$). This presupposes, of course, that M and N are at least of class C^r. We use 'smooth' synonymously with C^∞. The tangent bundle of a differentiable manifold M is denoted by TM. A C^r–*immersion* $f\colon M \to N$ is an $f \in C^r(M, N)$, $r \geq 1$, whose differential $Tf\colon TM \to TN$ is fibrewise injective. Recall that a manifold M is called *closed* if each component of M is compact and without boundary; it is called *open* if each component is noncompact or with nonempty boundary. Usually we shall implicitly assume that our manifolds are connected; in that case 'open' and 'closed' are complementary notions. The following theorem is commonly known as the Smale-Hirsch theorem. It was proved in the special case $M = S^m$, $N = \mathbb{R}^q$, $q \geq m + 1$, by S. Smale, and in its general form by M. Hirsch. There are other proofs due to R. Palais and V. Poénaru.

THEOREM 1.1 (Smale-Hirsch). *Let M and N be smooth manifolds. Assume that $\dim M < \dim N$ or that M is open. Then $f \in C^0(M, N)$ is homotopic to a smooth immersion of M into N if and only if f is covered by a fibrewise injective bundle map $F \in C^0(TM, TN)$.*

To say that f is covered by F means that we have a commutative diagram

$$\begin{array}{ccc} TM & \xrightarrow{F} & TN \\ \downarrow & & \downarrow \\ M & \xrightarrow{f} & N, \end{array}$$

where the vertical maps are the obvious projections.

Another typical application of the methods described in these lectures concerns the existence of symplectic structures. Recall that a *symplectic form* on a smooth manifold M of dimension $2n$ is a differential 2–form ω that is closed ($d\omega = 0$) and nondegenerate ($\omega^n \neq 0$, i.e. nowhere zero). Notice that the latter is a pointwise condition, that is, the nondegeneracy of ω on the fibre $T_x M$ of the tangent bundle only depends on the value of ω at $x \in M$.

THEOREM 1.2 (Gromov). *If M is open, then it admits a symplectic form if and only if it admits a nondegenerate 2–form.*

Thus, in both theorems we have a situation where an obviously necessary condition on the tangential level turns out to be sufficient for solving a geometric problem on a manifold. So-called h-principles (homotopy principles) deal with exactly this type of problems in the general context of differential equations or inequalities on manifolds. Sometimes they are just a convenient and neat way of formulating a result of this kind, but more often than not one can prove a general h-principle that allows one to deal with the analytical problems for a large class of geometric set-ups. This reduces the relevant geometric questions to obstruction theoretic problems that can be dealt with using methods from algebraic topology.

In other words, geometric problems that are governed by an h-principle typically admit more solutions (even parametric families) than the initial 'rigid' geometric set-up may suggest. This is what we call, on an informal level, 'flexibility' of solutions. In the context of the h-principle for open and invariant relations (defined below), the term 'flexibility' also takes on a more technical meaning.

ACKNOWLEDGEMENTS. These notes more or less faithfully reproduce a course of six lectures I presented in the Graduiertenkolleg 'Analysis, Geometrie und ihre Verbindung zu den Naturwissenschaften' (coordinated by H.-B. Rademacher) of Leipzig University between 12 and 20 December 2000. I thank Matthias Schwarz for inviting me to give this minicourse and for making my stay in Leipzig such a pleasant one. I am also grateful to Leipzig University and the Max-Planck Institute for Mathematics in the Sciences for their hospitality and support.

CHAPTER 2

Differential Relations and h-Principles

Differential relations are the generalisation to manifolds and fibre bundles of differential equations or inequalities. The key concept in the language that is necessary for formulating such a generalisation is that of a *jet*. We only present some of the basic definitions and concentrate on 1–jets. The textbooks [1] and [17] contain more (yet concise) information about jets in general. I discuss a few bundle-theoretic aspects that are not treated in these books. A comprehensive reference for the theory of jet bundles is [27].

Consider a smooth fibre bundle $p\colon E \to M$ over a smooth manifold M. A *section* of E is a map $\sigma\colon M \to E$ with $p \circ \sigma = \mathrm{id}_M$. We write $\Gamma^r(E)$ for the space of C^r–sections of E. We say that two local sections σ_1 and σ_2 defined in a neighbourhood of some point $x \in M$ are equivalent if in some (and hence any) description in terms of local coordinates on M and E they have the same derivatives up to order r. An equivalence class under this relation is called an r–*jet*. The jet defined by σ is denoted by $j_x^r \sigma$. In local coordinates, an r–jet $j_x^r \sigma$ has a canonical representative, namely, the order r Taylor polynomial of σ at x. We denote by E^r the space of r–jets of local C^r–sections of E.

Most of the examples we are going to consider can be phrased in terms of 1–jets. These can be described a little more explicitly. Two local sections σ_1 and σ_2 define the same 1–jet at $x \in M$ if and only if $\sigma_1(x) = \sigma_2(x)$ and $T_x \sigma_1 = T_x \sigma_2$. Thus, a 1–jet at x can be described as a pair
$$j_x^1 \sigma = (\widetilde{x}, L),$$
where $\widetilde{x} \in E$ is a point with $p(\widetilde{x}) = x$ and $L\colon T_x M \to T_{\widetilde{x}} E$ is a linear map satisfying
$$Tp \circ L = \mathrm{id}\colon T_x M \to T_x M,$$
since L has to correspond to the differential of a local section.

In the special case of a product bundle $E = M \times N$, we can naturally identify $\Gamma^r(E)$ with $C^r(M, N)$, and a 1–jet at $x \in M$ may be thought of as a pair (y, L') consisting of a point $y \in N$ and a linear map $L' \colon T_x M \to T_y N$.

There are natural fibrations $p^r \colon E^r \to M$ and $p_s^r \colon E^r \to E^s$ for $r \geq s \geq 0$ (where we set $E^0 = E$), satisfying $p \circ p_0^r = p^r$ and $p_t^s \circ p_s^r = p_t^r$ for $r \geq s \geq t \geq 0$. In the case $r = 1$ we can write these explicitly as

$$p^1(\widetilde{x}, L) = x \quad \text{and} \quad p_0^1(\widetilde{x}, L) = \widetilde{x}.$$

The fibre $(p^r)^{-1}(x)$ of E^r over $x \in M$ will be written as E_x^r.

In the following four lemmas we collect some basic observations about these fibrations.

LEMMA 2.1. *For any fibre bundle $p \colon E \to M$ with $\dim M = m$ and fibre F of dimension q, the jet bundle $p_0^1 \colon E^1 \to E$ is an affine bundle with fibre of dimension qm.*

Similar statements hold for all p_s^r.

PROOF. Let $U \subset M$ be an open subset over which the bundle E is trivial, and fix a trivialisation $E|U \cong U \times F$. Let $j_{x_0}^1 \sigma$ be the 1–jet at $x_0 \in U$ of a local section σ defined near x_0, where we think of σ as a map into F. Choose local coordinates $(x^1, \ldots, x^m)$ on U near x_0 and $(y^1, \ldots, y^q)$ on F near $\sigma(x_0)$. Write $\sigma^i = y^i \circ \sigma$. Then $j_{x_0}^1 \sigma = (\sigma(x_0), L')$, where $L' \colon T_{x_0} M \to T_{\sigma(x_0)} F$ is represented by the $(q \times m)$–matrix

$$(\sigma_j^i) := \left(\frac{\partial \sigma^i}{\partial x^j}(x_0) \right)_{1 \leq i \leq q,\, 1 \leq j \leq m}.$$

Since any Taylor polynomial in these local coordinates defines a jet, we see that the fibre of p_0^1 is given by the space $\mathcal{M}^{q,m} \cong \mathbb{R}^{qm}$ of $(q \times m)$–matrices.

It remains to show that coordinate transformations in U (near x_0) and F (near the point $\sigma(x_0)$) and changes of bundle trivialisations give rise to affine transformations of this fibre. This is clear for coordinate changes. In fact, here the transformation is given by pre- or postcomposing (σ_j^i) with the Jacobian of the coordinate change, which amounts to a linear transformation of $\mathbb{R}^{qm}$.

Given two local trivialisations $U_\alpha \times F$ and $U_\beta \times F$, they are related by a transition map

$$(U_\alpha \cap U_\beta) \times F \longrightarrow (U_\alpha \cap U_\beta) \times F$$
$$(x, y) \longmapsto (x, \phi(x, y)),$$

where ϕ is smooth and $\phi(x, .)$ is an element of the structure group of E for each x. So a section $(x, \sigma(x))$ transforms to $(x, \widetilde{\sigma}(x)) = (x, \phi(x, \sigma(x)))$. We compute in local coordinates, using summation convention; evaluation of the partial derivatives of $\sigma, \widetilde{\sigma}$ and ϕ at x_0 and $(x_0, \sigma(x_0))$, respectively, is understood:

$$\frac{\partial \widetilde{\sigma}^i}{\partial x^j} = \frac{\partial \phi^i}{\partial x^j} + \frac{\partial \phi^i}{\partial y^l} \frac{\partial \sigma^l}{\partial x^j}.$$

So the transformation $(\sigma_j^i) \rightsquigarrow (\widetilde{\sigma}_j^i)$ is an affine transformation of $\mathbb{R}^{qm}$. $\square$

LEMMA 2.2. *If* $p\colon E = M \times F \to M$ *is a product bundle, then* p_0^1 *is a vector bundle.*

PROOF. This is immediate from the considerations in the preceding proof, since in the case of a fixed global trivialisation $E = M \times F$ the local representations of the linear maps L' only differ by linear transformations. $\square$

LEMMA 2.3. *If* $p\colon E \to M$ *is a vector bundle, then so is the jet bundle* $p^1\colon E^1 \to M$.

PROOF. In local coordinates $(x^1, \ldots, x^m)$ on M and with *linear* coordinates $(y^1, \ldots, y^q)$ on $F \cong \mathbb{R}^q$, a 1–jet $j_x^1 \sigma$ is represented by

$$(\sigma^i, \sigma_j^i) \in \mathbb{R}^{q+qm}.$$

A coordinate transformation on M leaves the σ^i invariant and changes the σ_j^i by a linear transformation as in the proof of Lemma 2.1. A linear coordinate transformation of F changes (σ^i, σ_j^i) by a linear transformation of $\mathbb{R}^{q+qm}$.

Finally, we observe that different local trivialisations are related by transition maps $\phi(x, y)$ that are linear in y for fixed x, i.e.

$$\widetilde{\sigma}^i(x) = a_l^i(x) \sigma^l(x).$$

It follows that
$$(\widetilde{\sigma}^i, \widetilde{\sigma}^i_j) = \left(a^i_l \sigma^l, a^i_{l,j}\sigma^l + a^i_l \sigma^l_j\right),$$
which amounts to a linear transformation of $\mathbb{R}^{q+qm}$. $\square$

The key reason why we get a vector bundle and not just an affine bundle in the situations described in Lemmas 2.2 and 2.3 is the existence of a preferred reference section (defining the zero section in the jet bundle in question): In 2.2, the constant section; in 2.3, the zero section.

The next lemma will play a crucial role in the proof of Theorem 1.2. Write T^*M for the cotangent bundle of M and $\Lambda^2 T^* M$ for its second exterior power, so that the smooth sections of $\Lambda^2 T^* M$ are the differential 2–forms on M. With d we denote the usual exterior derivative on differential forms.

LEMMA 2.4. *Let $E = T^*M$. There is a commutative diagram*

$$\begin{array}{ccc} E^1 & \xrightarrow{\Delta} & \Lambda^2 T^* M \\ p^1 \downarrow & & \downarrow \\ M & \xrightarrow{\mathrm{id}} & M, \end{array}$$

with Δ a vector bundle epimorphism and an affine fibration, satisfying
$$\Delta j^1_{x_0} \alpha = (d\alpha)_{x_0}$$
for any local section α of E (i.e. differential 1–form) defined near the point $x_0 \in M$.

PROOF. Choose local coordinates $(x^1, \ldots, x^m)$ near a given point $x_0 \in M$. These determine local 1–forms $dx^1, \ldots, dx^m$ and hence a local trivialisation of E. In terms of this local trivialisation, the polynomial representative of the 1–jet of a local 1–form $\alpha = \alpha_i\, dx^i$ is of the form

$$j^1_{x_0}\alpha = (dx^1, \ldots, dx^m) \begin{pmatrix} a_1 \\ \vdots \\ a_m \end{pmatrix} + (dx^1, \ldots, dx^m) A \begin{pmatrix} x^1 - x^1_0 \\ \vdots \\ x^m - x^m_0 \end{pmatrix},$$

where $a_i = \alpha_i(x_0)$ and $A = (a_{ij}) = \left(\frac{\partial \alpha_i}{\partial x^j}(x_0)\right)$. We shall write
$$j^1_{x_0}\alpha = (a_i, a_{ij}) \in \mathbb{R}^{m+m^2}$$

for short. By the proof of Lemma 2.2, the vector bundle structure on $p^1\colon E^1 \to M$ is defined by the usual vector space structure on $\mathbb{R}^{m+m^2}$ under the described identification. Define

$$\Delta j^1_{x_0}\alpha = (dx^1,\ldots,dx^m)A\begin{pmatrix} dx^1 \\ \vdots \\ dx^m \end{pmatrix}$$

$$= \frac{\partial \alpha_i}{\partial x^j}(x_0)\,dx^i \wedge dx^j$$

$$= (d\alpha)_{x_0}.$$

Clearly Δ is well-defined and a vector bundle morphism covering the identity map on M. We have

$$\Delta^{-1}\Big(\sum_{i<j} b_{ij}\,dx^i \wedge dx^j|_{x_0}\Big) =$$

$$= \big\{(a_i, a_{ij}) \in \mathbb{R}^{m+m^2}\colon (a_{ij} - a_{ji}) = b_{ij} \text{ for } i<j\big\},$$

in particular Δ is surjective. We may choose a_i with $1 \leq i \leq m$ and a_{ij} with $1 \leq i \leq j \leq m$ as the free variables in the fibres of Δ. Computations completely analogous to those in the proof of Lemma 2.3 show that a change in local coordinates on M leads to an affine transformation of the coordinates in the fibres of Δ. $\square$

The attentive reader will have realised that continuous sections of E^r are not, in general, r–jets of C^r–sections of E (even locally). It is worth considering the difference in the simplest case, that of 1–jets. Let U be an open subset of M. Then a section $\sigma \in \Gamma^1(E|U)$ gives rise to a section $j^1\sigma \in \Gamma^0(E^1|U)$ defined by

$$j^1_x\sigma = (\sigma(x), T_x\sigma\colon T_xM \to T_{\sigma(x)}E).$$

An arbitrary section $\varphi \in \Gamma^0(E^1)$, on the other hand, is of the form

$$\varphi(x) = (\varphi_0(x), L_x),$$

where $\varphi_0 \in \Gamma^0(E)$ and $L_x\colon T_xM \to T_{\varphi_0(x)}E$ is a linear map whose composition with $T_{\varphi_0(x)}p$ gives the identity map on T_xM. Moreover, L_x is required to vary continuously with x. Thus, in general, φ_0 is not continuously differentiable and, even if it was, there is no need for L_x to equal $T_x\varphi_0$.

To give a more concrete example (that will appear again later on): For the trivial fibration $p\colon E = \mathbb{R} \times \mathbb{R}^3 \to \mathbb{R}$, elements of $\Gamma^0(E^1)$ can be interpreted as continuous curves in $\mathbb{R}^3$ with a continuously varying vector field along it, whereas the 1–jet of an element in $\Gamma^1(E)$ corresponds to a C^1–curve in $\mathbb{R}^3$ with its velocity vector field.

We now come to the basic notions in these lectures, due to M. Gromov [15].

DEFINITION 2.5. We say that $\varphi \in \Gamma^0(E^r)$ is **holonomic** if $\varphi = j^r\sigma$ for some $\sigma \in \Gamma^r(E)$. A **differential relation** (of order r) is a subset $\mathcal{R} \subset E^r$. A **solution** of $\mathcal{R}$ is a $\sigma \in \Gamma^r(E)$ with $j^r\sigma \in \Gamma^0(\mathcal{R})$. We say that $\mathcal{R}$ **satisfies the h-principle** (short for **homotopy principle**) if every continuous section of $\mathcal{R}$ is homotopic through sections of $\mathcal{R}$ to a holonomic section, i.e. a solution of $\mathcal{R}$.

There are considerably stronger notions of h-principles (parametric, relative, dense, ...), some of which we shall encounter presently. To give a first example of a relation satisfying the h-principle (in its weakest form), we may consider the trivial bundle $E = M \times N$ and the *immersion relation* $\mathcal{I} \subset E^1$ defined by requiring the fibre $\mathcal{I}_x = \mathcal{I} \cap E^1_x$ to consist of pairs (y, L) with $y \in N$ and L an injective linear map $T_xM \to T_yN$. Then the Smale-Hirsch theorem is seen to be equivalent to saying that $\mathcal{I}$ satisfies the h-principle.

Of course, the Smale-Hirsch theorem had been proved before the invention of the language of h-principles. The full strength of this language will become apparent when we see how it allows to prove h-principles for very general classes of relations, subsuming many older theorems of differential topology, but also generating a plethora of new applications.

In his book [15], Gromov describes three principal methods for proving h-principles:

1. Covering homotopy method (or the method of (micro-)flexible sheaves),
2. Convex integration,
3. Removal of singularities.

The first two of these I will describe in these lectures. As to the first one, I shall follow largely some old lecture notes of A. Haefliger [16] on Gromov's results, adding some additional details that the novice may find helpful, and indicating its relation with Gromov's later and slightly more sophisticated approach.

As to convex integration, I limit myself to describing the key idea of the method and illustrating a few examples. That way I hope to convey the ingenuity of the method without getting bogged down in technical details that would only help to obscure the argument. To get a working knowledge of this method one will obviously have to do some further reading in Gromov's book [15] or the monograph [28] by D. Spring.

While I was preparing these notes, Ya. Eliashberg and N. Mishachev have been developing an alternative approach to the h-principle [7], [8]. I have resisted the temptation to revise the present notes in the light of these new developments, but the occasional remark was included after conversations with Ya. Eliashberg, which I gratefully acknowledge.

CHAPTER 3

The h-Principle for open, invariant Relations

Following and expanding on the notes of Haefliger [16], this chapter will give a detailed proof of a general h-principle that includes as applications the theorems mentioned in the introduction, as well as several others that will be described below. The proof uses the covering homotopy method.

3.1. Open, invariant relations

Our first aim is to prepare for the precise formulation of the h-principle in question.

Denote by $\mathcal{D}(M)$ the pseudogroup of all local diffeomorphisms of a smooth manifold M. That is, $\mathcal{D}(M)$ consists of all diffeomorphisms $f\colon U \to V$ between arbitrary open subsets $U, V \subset M$. Then $\mathcal{D}(M)$ has the following properties, which constitute the axioms of a *pseudogroup of local diffeomorphisms*.

(i) The restriction of $f \in \mathcal{D}(M)$ to any open subset of the domain U of f is in $\mathcal{D}(M)$.
(ii) If $U = \cup U_i$ and f is a diffeomorphism of U onto an open set V, then f belongs to $\mathcal{D}(M)$ provided that each $f|U_i$ belongs to $\mathcal{D}(M)$.
(iii) For every open subset $U \subset M$, the identity map of U onto U belongs to $\mathcal{D}(M)$.
(iv) With $f \in \mathcal{D}(M)$ we also have $f^{-1} \in \mathcal{D}(M)$.
(v) If $f_i\colon U_i \to V_i$, $i = 1, 2$ belong to $\mathcal{D}(M)$ and $V_2 \subset U_1$, then $f_1 \circ f_2$ belongs to $\mathcal{D}(M)$.

Of course, all these statements are obvious in the case of $\mathcal{D}(M)$. One may similarly consider pseudogroups of local diffeomorphisms preserving an additional geometric structure on M, in which case the axioms may be less trivially satisfied.

As previously, we consider a smooth fibre bundle $p\colon E \to M$.

DEFINITION 3.1. A map $\Phi\colon \mathcal{D}(M) \to \mathcal{D}(E)$ is a **continuous extension** of $\mathcal{D}(M)$ to $\mathcal{D}(E)$ if the following conditions hold:

(i) For any $f\colon U \to V$ in $\mathcal{D}(M)$ the corresponding $\Phi(f)$ is a diffeomorphism of $p^{-1}(U)$ onto $p^{-1}(V)$ and we have a commutative diagram

$$\begin{array}{ccc} p^{-1}(U) & \xrightarrow{\Phi(f)} & p^{-1}(V) \\ {\scriptstyle p}\downarrow & & \downarrow{\scriptstyle p} \\ U & \xrightarrow{f} & V. \end{array}$$

(ii) $\Phi(\mathrm{id}_U) = \mathrm{id}_{p^{-1}(U)}$.

(iii) $\Phi(f \circ g) = \Phi(f) \circ \Phi(g)$ whenever $f \circ g$ is defined.

(iv) For any open set $U \subset M$ the map

$$\Phi\colon \mathrm{Diff}\,(U) \longrightarrow \mathrm{Diff}\,(p^{-1}(U))$$

between the groups of self-diffeomorphisms of the respective open subsets is continuous with respect to the weak C^∞-topology (or compact-open C^∞-topology) on these groups.

Cf. [**17**] for the definition of the weak C^r-topology. Convergence with respect to this topology, for r finite, means uniform convergence on compact sets of all derivatives up to order r; the C^∞-topology is the union of the C^r-topologies for r finite, induced by the inclusion of the space of C^∞-maps in the spaces of C^r-maps.

Here are three examples of continuous extensions:

(1) $E = M \times N$, $\Phi(f) = f \times \mathrm{id}$.

(2) $E = TM$, $\Phi(f) = Tf$.

(3) Given a continuous extension $\Phi\colon \mathcal{D}(M) \to \mathcal{D}(E)$, a continuous extension $\Phi^r\colon \mathcal{D}(M) \to \mathcal{D}(E^r)$ is defined as follows: For a diffeomorphism $f\colon U \xrightarrow{\cong} V$, set

$$\begin{array}{rccc} \Phi^r(f)\colon & (p^r)^{-1}(U) & \longrightarrow & (p^r)^{-1}(V) \\ & j^r_x \sigma & \longmapsto & j^r_{f(x)}(\Phi(f) \circ \sigma \circ f^{-1}). \end{array}$$

This is the natural definition to make, as is clear from the commutative diagram in part (i) of Definition 3.1.

Only the verification of the continuity condition (iv) requires a small argument, which I leave to the reader.

3.2. Statement of the theorem

We have now discussed all technical notions necessary to state our first general h-principle, that for open and invariant relations. In order to simplify the notation in the theorem and its proof, we introduce, for a given differential relation $\mathcal{R} \subset E^r$, the following spaces. As in the example of the immersion relation, we write $\mathcal{R}_x = \mathcal{R} \cap E^r_x$ for the fibre of $\mathcal{R}$ over $x \in M$. Furthermore, we define

$$\Gamma \mathcal{R} = \{\varphi \in \Gamma^0(E^r) \colon \varphi(x) \in \mathcal{R}_x \ \forall x \in M\}$$

and

$$\Gamma_0 E = \{\sigma \in \Gamma^r(E) \colon j^r\sigma \in \Gamma\mathcal{R}\}.$$

These spaces will be equipped with the weak (C^0 resp. C^r) topology. Notice that with the weak topologies on $\Gamma^r E$ and $\Gamma^0 E^r$ the r–jet map

$$j^r \colon \Gamma^r E \longrightarrow \Gamma^0 E^r$$

is continuous, and it restricts to a continuous map

$$j^r \colon \Gamma_0 E \longrightarrow \Gamma\mathcal{R}.$$

DEFINITION 3.2. Let $p \colon E \to M$ be a smooth fibre bundle and $\mathcal{R} \subset E^r$ an open subbundle. If there is a continuous extension $\Phi \colon \mathcal{D}(M) \to \mathcal{D}(E)$ such that $\mathcal{R}$ is invariant under Φ^r (as defined in example (3) above), we call $\mathcal{R}$ an **open, invariant relation**.

THEOREM 3.3 (Gromov). *Let M be an open manifold and $\mathcal{R} \subset E^r$ an open, invariant relation over M. Then*

$$j^r \colon \Gamma_0 E \longrightarrow \Gamma\mathcal{R}$$

is a weak homotopy equivalence (w.h.e.).

This means that under the assumptions of the theorem, j^r induces a one-to-one correspondence between the pathwise connected components of $\Gamma_0 E$ and $\Gamma\mathcal{R}$, and the induced map on homotopy groups

$$j^r_\# \colon \pi_i(\Gamma_0 E, \sigma) \longrightarrow \pi_i(\Gamma\mathcal{R}, j^r\sigma)$$

is a group isomorphism for each $i \geq 1$ and any $\sigma \in \Gamma_0 E$.

Notice that surjectivity on π_0 (i.e. the set of pathwise connected components) is what in Definition 2.5 was referred to as 'h-principle'.

Injectivity on π_0 means that two solutions of $\mathcal{R}$ whose r–jets are homotopic as sections of $\mathcal{R}$ are actually homotopic as solutions. To give a concrete example: Two immersions $M \to N$ (with M and N as in Theorem 1.1) whose differentials are homotopic as fibrewise injective bundle maps are homotopic as immersions; see Section 3.3 below. This injectivity on π_0 is also called the 1–**parametric h-principle**. The statement that j^r is a w.h.e. is referred to as the **(multi-)parametric h-principle**.

By a simple example one can illustrate that Theorem 3.3 fails, in general, for closed manifolds: Let $M = S^1$ and E the trivial bundle $E = S^1 \times \mathbb{R}$. Then, since the tangent bundle of S^1 is trivial, we have

$$E^1 = S^1 \times \mathbb{R} \times L(\mathbb{R}, \mathbb{R}) \cong S^1 \times \mathbb{R} \times \mathbb{R},$$

where $L(\mathbb{R}, \mathbb{R})$ denotes the space of linear maps $\mathbb{R} \to \mathbb{R}$. The immersion relation

$$\mathcal{I} = S^1 \times \mathbb{R} \times \operatorname{GL}(1, \mathbb{R}) \cong S^1 \times \mathbb{R} \times (\mathbb{R} - \{0\})$$

is open, invariant, and non-empty, so $\Gamma \mathcal{I} \neq \emptyset$. But $\Gamma_0 E$ is the space of C^1–immersions of S^1 in $\mathbb{R}$, which is empty.

3.3. Applications

We can now deduce as corollaries of Gromov's Theorem 3.3 the two theorems stated in the introduction. We also provide additional more concrete examples for the Smale-Hirsch immersion theorem, in particular the celebrated Smale paradox concerning the eversion of S^2 in $\mathbb{R}^3$, a proof of the Phillips submersion thorem, as well as applications to symplectic, contact, and Riemannian geometry.

3.3.1. The Smale-Hirsch immersion theorem. Let M and N be smooth manifolds, with M open. Write $\mathcal{I}mm(M, N)$ for the space of C^1–immersions $M \to N$ with the weak C^1–topology and $\mathcal{M}on(TM, TN)$ for the space of continuous and fibrewise injective bundle maps $TM \to TN$ with the weak C^0–topology. We have already seen after Definition 2.5 how to translate this into a differential relation $\mathcal{I} \subset E^1$, where $E = M \times N$. This relation is open and invariant,

3.3. APPLICATIONS

so from Theorem 3.3 we conclude that

$$\mathcal{I}mm(M,N) \longrightarrow \mathcal{M}on(TM,TN)$$
$$f \longmapsto Tf$$

is a w.h.e. In particular, this means that a map $f \in C^0(M,N)$ is homotopic to a C^1–immersion f_0 if and only if f is covered by a bundle map $F \in \mathcal{M}on(TM,TN)$, and f_0 can be chosen in such a way that F and Tf_0 are homotopic in $\mathcal{M}on(TM,TN)$. Such a C^1–immersion f_0 can be approximated in the strong C^1–topology by a smooth map $f_1 \colon M \to N$, see [**17**, Thm. 2.2.6]. If f_1 is sufficiently close to f_0, then it will still be an immersion (by the openness of $\mathcal{I}mm(M,N)$ in $C^1(M,N)$ with the strong C^1–topology, see [**17**, Thm. 2.1.1]), and in fact homotopic to f_0 through immersions (or what is called *regularly homotopic* to f_0); cf. Section 3.5 for the type of argument necessary to fill in the details of this last step[1]. This proves Theorem 1.1 for open M.

As a particular instance of this result we have the *Smale paradox*. Let $M = S^n \times \mathbb{R}$ and $N = \mathbb{R}^{n+1}$. Then $TM \cong (S^n \times \mathbb{R}) \times \mathbb{R}^{n+1}$. By fixing trivialisations and fibre orientations of TM and TN we can identify $F \in \mathcal{M}on^+(TM,TN)$, where the '+' denotes preservation of fibre orientation, with a map

$$S^n \times \mathbb{R} \longrightarrow \mathbb{R}^{n+1} \times \mathrm{GL}^+(n+1,\mathbb{R}).$$

Furthermore, the set $\pi_0(\mathcal{M}on^+(TM,TN))$ corresponds to the set of homotopy classes of such maps, which equals $\pi_n(\mathrm{GL}^+(n+1,\mathbb{R}))$. To compute that homotopy group, we recall the fact that any matrix $A \in GL^+(n+1,\mathbb{R})$ has a unique polar decomposition $A = PR$ with P symmetric and positive definite, and $R \in \mathrm{SO}(n+1)$. Since the space of symmetric, positive definite matrices is convex, and hence contractible, we conclude

$$\pi_0(\mathcal{M}on^+(TM,TN)) = \pi_n(\mathrm{SO}(n+1)).$$

---

[1]This step can be avoided if instead one works with the spaces of smooth immersions and smooth fibrewise injective bundle maps (still with the weak C^1– resp. C^0–topology). This would be the approach taken in [**26**], for instance, in the corresponding problem for submersions. In the present notes I follow Gromov's approach, imposing only the minimal differentiability assumption.

Moreover, an immersed S^n in $\mathbb{R}^{n+1}$ always has an orientable and hence trivial normal bundle, so there is a homotopically unique way to extend an immersion $S^n \to \mathbb{R}^{n+1}$ to an orientation preserving immersion $S^n \times \mathbb{R} \to \mathbb{R}^{n+1}$ (cf. the proof of uniqueness of tubular neighbourhoods in [17]). In conclusion we have the following result:

PROPOSITION 3.4. *Regular homotopy classes of immersions* $S^n \to \mathbb{R}^{n+1}$ *are classified by* $\pi_n(\mathrm{SO}(n+1))$.

Thus, immersions $S^1 \to \mathbb{R}^2$ are classified by $\pi_1(\mathrm{SO}(2)) \cong \mathbb{Z}$. An explicit isomorphism

$$\pi_0(\mathcal{I}mm(S^1, \mathbb{R}^2)) \xrightarrow{\cong} \mathbb{Z}$$

can be defined by associating to an immersion the rotation number of its tangent vectors with respect to a reference frame in $\mathbb{R}^2$. This is a classical result from the 1930's due to H. Whitney [32], cf. [1] for an exposition along the lines of Whitney's original paper. In Chapter 4 we shall discuss Whitney's result in the context of convex integration theory.

Any compact Lie group has vanishing second homotopy group, so we have $\pi_2(\mathrm{SO}(3)) = 0$ (which can of course be seen more easily from an explicit identification of $\mathrm{SO}(3)$ with projective 3–space). This implies that all immersions $S^2 \to \mathbb{R}^3$ are regularly homotopic. In particular, we can consider the standard inclusion $\iota\colon S^2 \to \mathbb{R}^3$ and the embedding $\iota \circ \alpha \colon S^2 \to \mathbb{R}^3$, where α is the antipodal map of S^2. Assume that orientations have been chosen in such a way that in the extension of ι to an immersion $S^2 \times \mathbb{R} \to \mathbb{R}^3$ the positive $\mathbb{R}$–direction corresponds to the *outward* normal of S^2 in $\mathbb{R}^3$. Since α is orientation reversing, this implies that in the extension of $\iota \circ \alpha$ to an immersion $S^2 \times \mathbb{R} \to \mathbb{R}^3$ the positive $\mathbb{R}$–direction has to correspond to the *inward* normal. So the regular homotopy from ι to $\iota \circ \alpha$ reverses the normal direction, i.e. it 'turns the sphere inside out'. This phenomenon is known as the *Smale paradox*.

To some extent, this can be visualised as follows. Start with an immersion of $\mathbb{R}P^2$ in $\mathbb{R}^3$ (e.g. the famous Boy surface). Since $\mathbb{R}P^2$ is nonorientable, its normal bundle in $\mathbb{R}^3$ is nontrivial, so the boundary of a tubular neighbourhood $N(\mathbb{R}P^2)$ defines an immersion of S^2

in $\mathbb{R}^3$. This immersion can obviously be turned inside out by a regular homotopy of $\partial N(\mathbb{R}P^2)$ that exchanges the two ends of each fibre of $N(\mathbb{R}P^2) \to \mathbb{R}P^2$ by shifting them along that fibre. However, it is of course not clear *a priori* that the immersion $S^2 \to \partial N(\mathbb{R}P^2) \subset \mathbb{R}^3$ is regularly homotopic to the standard embedding $S^2 \subset \mathbb{R}^3$.

For nice illustrations of the Smale paradox see [25] and [9].

In order to prove Theorem 1.1 for closed manifolds M with $m = \dim M < \dim N = n$, we need a little bundle theory. Write $p_M \colon TM \to M$ and $p_N \colon TN \to N$ for the bundle projections. Given a map $f \in C^0(M, N)$, there is an induced *pull-back bundle* over M,

$$f^*TN = \{(x, Y) \in M \times TN \colon f(x) = p_N(Y)\}.$$

The bundle structure is given by the obvious map $f^*TN \to M$, and the equally obvious map $f^*TN \to TN$ yields the commutative diagram

$$\begin{array}{ccc} f^*TN & \longrightarrow & TN \\ \downarrow & & \downarrow {\scriptstyle p_N} \\ M & \xrightarrow{f} & N. \end{array}$$

If f is an immersion (at least C^1), one can define its *normal bundle* by

$$\nu_f = f^*TN/TM,$$

where TM is regarded as a subbundle of f^*TN via the bundle inclusion

$$\begin{array}{ccc} TM & \longrightarrow & f^*TN \\ X & \longmapsto & (p_M(X), Tf(X)). \end{array}$$

Observe that regularly homotopic immersions f_0, f_1 have isomorphic normal bundles ν_{f_0}, ν_{f_1} (everything needed to prove this statement is contained in [18, Section 3.4], for instance). This suggests the following definitions. Fix an $(n - m)$–dimensional vector bundle ν over M. Set

$$\mathcal{I}mm_\nu(M, N) = \{f \in \mathcal{I}mm(M, N) \colon \nu_f \cong \nu\}$$

and

$$\mathcal{M}on_\nu(TM, TN) = \{F \in \mathcal{M}on(TM, TN) \colon \overline{F}^*TN/TM \cong \nu\};$$

here $\overline{F} \colon M \to N$ denotes the map induced by $F \colon TM \to TN$, and TM is regarded as a subbundle of $\overline{F}^*TN$ by means of the bundle inclusion $X \mapsto (p_M(X), F(X))$. Write $E(\nu)$ for the total space of ν

(this is an open manifold) and $\pi\colon E(\nu) \to M$ for the bundle projection. Observe that $T(E(\nu)) \cong \pi^*(TM \oplus \nu)$. This leads to the following diagram, where the vertical maps are induced by π and (on the right) the projection $TM \oplus \nu \to TM$:

$$\begin{array}{ccc} \mathcal{I}mm(E(\nu), N) & \xrightarrow{j^1} & \mathcal{M}on(\pi^*(TM \oplus \nu), TN) \\ \downarrow & & \downarrow \\ \mathcal{I}mm_\nu(M, N) & \xrightarrow{j^1} & \mathcal{M}on_\nu(TM, TN). \end{array}$$

One can now show that the vertical maps in this diagram are Serre fibrations (see e.g. [2] for this and any other standard notion of topology whose definition I omit in the present notes). Together with Lemma 3.15 below and the observation that the induced map on fibres is a w.h.e., one concludes that the 1–jet map at the bottom of the diagram is also a w.h.e.

To keep matters simple, we prove directly the weaker claim made in Theorem 1.1, i.e. surjectivity on π_0. Thus, let $f\colon M \to N$ be given and assume that it is covered by an $F \in \mathcal{M}on(TM, TN)$. With ν defined by f^*TN/TM we have $F \in \mathcal{M}on_\nu(TM, TN)$. This F easily lifts to an $\widetilde{F} \in \mathcal{M}on(\pi^*(TM \oplus \nu), TN)$. By the Smale-Hirsch theorem for open manifolds, we find an $\widetilde{f_1} \in \mathcal{I}mm(E(\nu), N)$ with $T\widetilde{f_1}$ homotopic to $\widetilde{F}$ in $\mathcal{M}on(\pi^*(TM \oplus \nu), TN)$. The restriction $f_1 = \widetilde{f_1}|M$ is a C^1–immersion homotopic to f.

A particular instance of the Smale-Hirsch theorem says that an m–dimensional *parallelisable* manifold (i.e. with trivial tangent bundle) admits an immersion into $\mathbb{R}^{m+1}$ or, if M is open, even into $\mathbb{R}^m$. In general, such immersions are well-nigh impossible to visualise. An explicit immersion, due to J. Milnor, of a punctured m–torus in $\mathbb{R}^m$ can be found in [19, pp. 47–49].

3.3.2. The Phillips submersion theorem. Recall that a map $f \in C^r(M, N)$, $r \geq 1$, is called a C^r–*submersion* if its differential $Tf\colon TM \to TN$ is fibrewise surjective. (It is not required that f be surjective.) We write $\mathcal{S}ub(M, N)$ for the space of C^1–submersions (with the weak C^1–topology) and $\mathcal{E}pi(TM, TN)$ for the space of continuous fibrewise surjective bundle maps $TM \to TN$ (with the weak

C^0–topology). The translation into a differential relation is now completely analogous to the case of immersions: We take $E = M \times N$ and define the submersion relation $\mathcal{S} \subset E^1$ by

$$\mathcal{S}_x = \{(y, L) \colon y \in N,\ L \colon T_xM \to T_yN \text{ surjective}\}.$$

This relation satisfies the conditions of Theorem 3.3, so we obtain the Phillips submersion theorem [**26**]:

THEOREM 3.5 (Phillips). *If M is open, then the map*

$$\begin{array}{rcl} \mathcal{S}ub(M,N) & \longrightarrow & \mathcal{E}pi(TM,TN) \\ f & \longmapsto & Tf \end{array}$$

is a w.h.e.

For $N = \mathbb{R}^n$ and $f = (f_1, \ldots, f_n)$, we can identify the differential Tf (of rank n) with the n–coframe $(df_1, \ldots, df_n)$. Using an auxiliary Riemannian metric $\langle .,. \rangle$ on M we can pass to the n–frame $(\nabla f_1, \ldots, \nabla f_n)$, where the gradient ∇ is defined by $\langle \nabla f_i, . \rangle = df_i(.)$. Write $\Gamma(V_n(TM))$ for the space of continuous n–frames on M.

COROLLARY 3.6. *If M is open, then the map*

$$\begin{array}{rcl} \mathcal{S}ub(M, \mathbb{R}^n) & \longrightarrow & \Gamma(V_n(TM)) \\ f & \longmapsto & (\nabla f_1, \ldots, \nabla f_n) \end{array}$$

is a w.h.e.

Here is a nice illustration of this corollary, also due to A. Phillips. Consider submersions of the open annulus $S^1 \times \mathbb{R}$ in $\mathbb{R}$. We may think of $S^1 \times \mathbb{R}$ as the open subset

$$\{(x,y) \in \mathbb{R}^2 \colon 1 < x^2 + y^2 < 2\} \subset \mathbb{R}^2$$

and define the gradient with respect to the standard metric on $\mathbb{R}^2$. We compute

$$\begin{aligned} \pi_0(\mathcal{S}ub(S^1 \times \mathbb{R}, \mathbb{R})) & = \pi_0 \Gamma(V_1(T(S^1 \times \mathbb{R}))) \\ & = \pi_0 \Gamma(S^1 \times \mathbb{R} \times (\mathbb{R}^2 - \{0\})) \\ & = \pi_1(S^1) \\ & = \mathbb{Z}. \end{aligned}$$

In the second line we have used the parallelisability of $S^1 \times \mathbb{R}$; the total equality represents a set bijection (not a group isomorphism: π_0 is only a set). This means that a submersion $S^1 \times \mathbb{R} \to \mathbb{R}$ is characterised up to regular homotopy (i.e. homotopy through submersions) by the winding number of its gradient vector field with respect to $\partial/\partial x$, say. For the submersion $S^1 \times \mathbb{R} \to \mathbb{R}$ obtained by first immersing $S^1 \times \mathbb{R}$ into $\mathbb{R}^2$ as a thickened figure '∞' and then projecting onto the x–axis that winding number equals 1. The same is true for the submersion given (in polar coordinates on $\mathbb{R}^2$) by $(r, \theta) \mapsto r$. It is a nice little exercise to show that those two submersions are regularly homotopic. (Hint: On the annulus, draw the level lines of the first submersion. Realise the same level lines as level lines of the function r by deforming $S^1 \times \mathbb{R} \subset \mathbb{R}^2$. See [26].) In a similar vein, one has the striking consequence, pointed out to me by V. Ginzburg, that the functions r and $-r$ on $\mathbb{R}^2 - \{0\}$ can be connected by a family of functions with nonvanishing gradient.

3.3.3. Symplectic and contact structures. We are now going to prove Theorem 1.2 from the introduction. I shall also discuss some related results in symplectic and contact geometry.

Let M be a smooth manifold of dimension $2n$ and let $E = T^*M$ be the cotangent bundle of M. Let $\Delta \colon E^1 \to \Lambda^2 T^*M$ be the affine fibration described in Lemma 2.4. Define a 1–jet relation $\mathcal{R}$ by

$$\mathcal{R} = \{u \in E^1 \colon (\Delta u)^n \neq 0\}.$$

Since Δ is an affine fibration (and in particular a homotopy equivalence), we see that $\Gamma \mathcal{R}$ is homotopy equivalent to the space

$$\{\beta \in \Gamma^0(\Lambda^2 T^*M) \colon \beta^n \neq 0\}$$

of nondegenerate (continuous) 2–forms. The relation $\mathcal{R}$ is open and invariant, so provided M is open we conclude from Theorem 3.3 that the existence of a nondegenerate 2–form implies $\pi_0(\Gamma_0 E) \neq 0$, i.e. there is a C^1–section α of T^*M such that

$$(d\alpha)^n = (\Delta j^1 \alpha)^n \neq 0.$$

An exact symplectic form $\omega = d\alpha'$ is found by C^1–approximating α by a smooth 1–form α'. This proves Theorem 1.2.

Here is a slight variation and strengthening of that theorem. Recall that an *almost complex structure* J is a vector bundle isomorphisms

$J\colon TM \to TM$ (covering the identity map) with $J_x^2 = -\mathrm{id}\colon T_xM \to T_xM$ for all $x \in M$. The space of nondegenerate 2–forms on M is in fact homotopy equivalent to the space of almost complex structures. The reason for this is that to a given β one can find a *compatible* J in the sense that $\beta(X, JY)$ defines a J–invariant Riemannian metric, and the space of such β–compatible J is contractible, see [**24**, Prop. 4.1]. Conversely, starting from an almost complex structure J and an arbitrary Riemannian metric $\langle .,.\rangle$, a nondegenerate 2–form β compatible with J is defined by

$$\beta(X,Y) = \langle Y, JX\rangle - \langle X, JY\rangle.$$

The space of J–compatible β is convex (since the space of metrics is) and hence contractible. More succinctly: The symplectic group $\mathrm{Sp}(2n,\mathbb{R})$ has the subgroup $\mathrm{U}(n)$ as a deformation retract.

Moreover, given a cohomology class $a \in H^2(M;\mathbb{R})$ represented by a closed 2–form λ, the 2–form $\lambda + Kd\alpha'$, (with α' from the proof of Theorem 1.2 above and $K \in \mathbb{R}$) still represents the class a, and it will be a symplectic form for K sufficiently large. This proves the following:

COROLLARY 3.7. *Let M be open, $a \in H^2(M;\mathbb{R})$, and J an almost complex structure on M. Then M admits a symplectic form ω with $[\omega] = a$ and with compatible almost complex structure homotopic to J.*

The odd-dimensional analogue of a symplectic form is a *contact form*: a differential 1–form α on a $(2n+1)$–dimensional manifold M satisfying the nondegeneracy (better: maximal nonintegrability) condition $\alpha \wedge (d\alpha)^n \neq 0$. The underlying tangential structure (i.e. the analogue of the nondegenerate 2–form resp. almost complex structure in the symplectic case) is a pair (β_1, β_2) consisting of a 1–form β_1 and a 2–form β_2 satisfying the condition $\beta_1 \wedge \beta_2^n \neq 0$ resp. a cooriented codimension 1 subbundle of TM (the kernel of β_1) with an almost complex structure (compatible with $\beta_2|_{\ker \beta_1}$). As in the symplectic case, these two descriptions of the tangential structure amount, up to homotopy, to the same thing. Such a structure is called an *almost contact structure*. For instance, an almost contact structure on a 3–manifold is simply an oriented and cooriented 2–plane field.

The relevant differential relation is now (still with $E = T^*M$)
$$\mathcal{R} = \{u \in E^1 \colon (p_0^1 u) \wedge (\Delta u)^n \neq 0\}.$$
Then Theorem 3.3 applies to show the following:

THEOREM 3.8. *On open odd-dimensional manifolds, the 1–jet map on differential 1–forms induces a w.h.e. between the spaces of contact forms and almost contact structures.*

Both in the symplectic and contact case, the h–principle fails, in general, on closed manifolds. In the symplectic case it is clear that the described approach cannot work, for it produced an *exact* symplectic form, which violates the cohomological condition $[\omega]^n \neq 0$ on a closed manifold. But this is more than just the failure of one particular method. For instance, S^6 is a manifold admitting an almost complex structure but no symplectic form, again for the same cohomological reason. So there can be no general h-principle for symplectic forms on closed manifolds. Gromov's theory of pseudoholomorphic curves and the results of C. Taubes in Seiberg-Witten theory, specifically the nonexistence of any symplectic form on $\mathbb{C}P^2 \# \mathbb{C}P^2 \# \mathbb{C}P^2$, imply that there is no h-principle for closed symplectic 4–manifolds, even with an extra cohomological condition. In higher dimensions this is an open question.

In the contact case, the situation is less clear. For instance, in dimension 3 Eliashberg introduced a dichotomy between so-called *tight* and *overtwisted* contact structures, and showed that the latter satisfy the parametric h-principle [6]. Also, contact structures on simply connected 5–manifolds satisfy the h-principle, see [10] (but not even the 1–parametric h–principle: on S^5 there are many nonhomotopic contact structures with the same underlying almost contact structure, cf. [11], [30]).

The proof of that last result in [10] makes essential use of the h–principle for *Legendre immersions*, due to T. Duchamp [5] and Gromov. These are immersions f of an n–dimensional manifold L in a $(2n+1)$–dimensional contact manifold (M, α) such that $T_x f(T_x L) \subset \ker \alpha_{f(x)}$ for all $x \in L$. There are several related concepts (isotropic immersions, contact immersions; or in the symplectic case: isotropic, Lagrange, symplectic immersions), all of which satisfy some form of h–principle under suitable conditions, see Gromov's book [**15**] and the forthcoming one by Eliashberg and Mishachev [**8**]. The references [**21**], [**3**] and [**4**] may also be useful.

3.3.4. Riemannian metrics. Restrictions on the curvature of a Riemannian metric can be formulated as differential relations in the space of 2–jets of local metrics. The simplest curvature restrictions are those requiring the sectional curvature (Sec), Ricci curvature (Ric) or scalar curvature (Scal) to be positive or negative. These are open, invariant relations, so Theorem 3.3 gives:

THEOREM 3.9. *On any open manifold, any of the curvature relations*
$$\text{Sec, Ric, Scal} \lessgtr 0$$
satisfies the parametric h-principle.

The situation changes drastically when a completeness constraint on the metric is introduced, thus in particular if we consider closed manifolds. For any of the described relations $\mathcal{R}$, the space $\Gamma\mathcal{R}$ turns out to be nonempty, but there are closed manifolds without any metrics satisfying Sec $\lessgtr 0$, Ric > 0 or Scal > 0, so for these relations the h-principle fails on closed manifolds. The following result of Lohkamp [**22**] is therefore quite striking.

THEOREM 3.10 (Lohkamp). *The relations* Ric < 0 *and* Scal < 0 *satisfy the (C^0-dense, parametric) h-principle.*

Here 'C^0-dense' means that, given a metric on a manifold M and any open neighbourhood of that metric in $\Gamma(\odot^2 T^*M)$ with the C^0–topology, a metric satisfying the relation Ric < 0 (or the weaker one Scal < 0) can be found in that neighbourhood. This is false, in general, for other curvature restrictions. For instance, the Bishop comparison theorem implies that a negatively curved metric (Sec < 0) can under no circumstances be C^0–approximated by a metric with Ric > 0.

Just as striking as the stated h-principle for Ricci negative Riemannian metrics is the following h-principle for isometric immersions, that is, immersions preserving the lengths of tangent vectors (so the immersion being C^1 is sufficient for the notion of 'isometric'). In the form stated here it is due to Gromov; essentially it belongs to J. Nash and N. Kuiper.

THEOREM 3.11 (Nash-Kuiper). *Let M be a Riemannian manifold of dimension m and let $\mathbb{R}^q$ be equipped with its standard Euclidean structure. For $q > m$, isometric C^1-immersions satisfy the parametric h-principle.*

The theorem remains true for arbitrary convex subsets of $\mathbb{R}^q$, which has the rather counterintuitive consequence that, for instance, the unit sphere $S^n \subset \mathbb{R}^{n+1}$ admits an isometric C^1-immersion into an arbitrarily small ball in $\mathbb{R}^{n+1}$. By contrast, any C^2-immersion of S^n is congruent to the unit sphere by classical rigidity results.

Isometric C^1-immersions translate into a non-open differential relation in a 1–jet bundle, so the methods discussed thus far do not apply to prove this difficult theorem. The Nash-Kuiper theorem is proved in Gromov's book [15] by the method of convex integration (see also the book by Eliashberg and Mishachev [8], where the argument is perhaps a little more direct). Convex integration will be discussed in Chapter 4 of the present notes.

3.4. Proof of the theorem

Let M be an open m–dimensional differential manifold. Then there exists a proper Morse function $f\colon M \to [0, \infty)$ with all critical points $a_1, a_2, \ldots$ of index less than m, exactly one critical point of index 0, and such that the sequence of critical values $c_i = f(a_i)$ is strictly increasing (and the indices of the corresponding critical points are increasing). A textbook reference for these facts is [20], see in particular Theorems VII 1.1 and VII 6.1. This means that in suitable local coordinates $(x_1, \ldots, x_m)$ around a_i the function f takes the form

$$f = c_i - x_1^2 - \ldots - x_k^2 + x_{k+1}^2 + \ldots + x_m^2$$

with $k = k_i < m$. Define

$$M_i = f^{-1}[0, c_i + \varepsilon_i]$$

for some ε_i satisfying $0 < \varepsilon_i < c_{i+1} - c_i$, see Figure 3.1.

Now define

$$W_i = \{x_1^2 + \ldots + x_k^2 < \delta_i\} \cap M_i$$

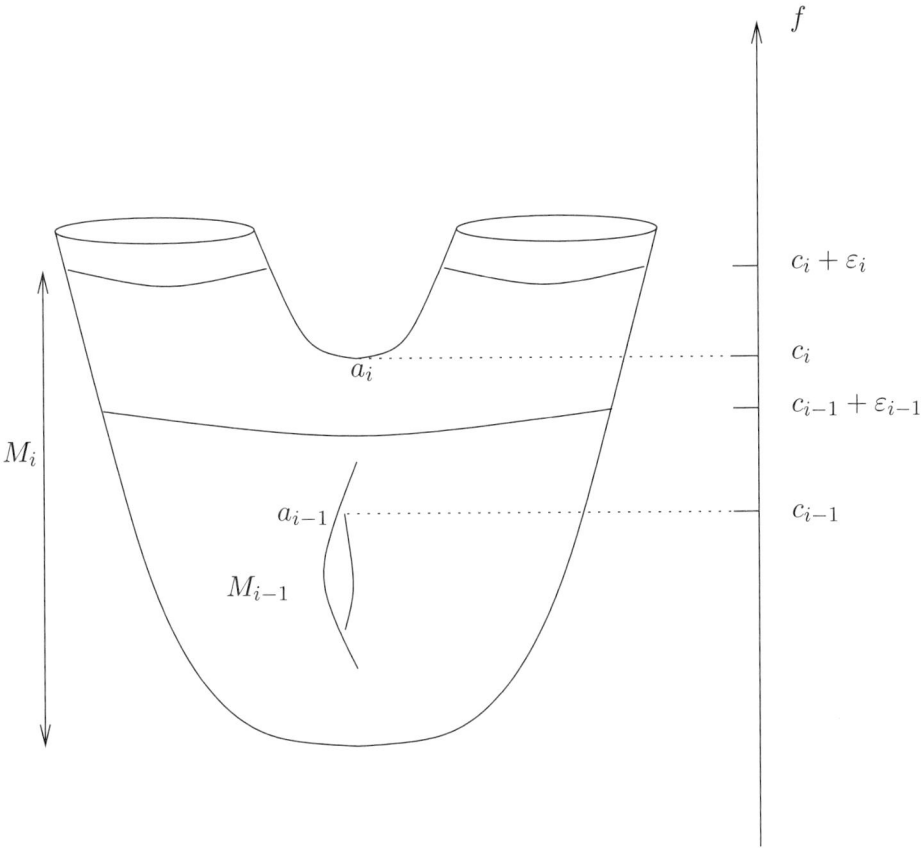

FIGURE 3.1. Exhaustion of M by the M_i.

for some small $\delta_i > 0$, and set $M'_{i-1} = M_i - W_i$, see Figure 3.2.

Observe that W_i is a diffeomorphic copy of $\text{Int}(D^k) \times D^{m-k}$. The manifold M'_i has a corner $S^{k-1} \times S^{m-k-1}$, and it may be thought of as a union of M_{i-1} with a collarlike neighbourhood in $\partial M_{i-1} \times [0, 1]$. Moreover, we shall consider a collarlike neighbourhood A_i of W_i in M_i, and fix identifications

$$W_i \cong \text{Int}(D^k_{1/2}) \times D^{m-k} \subset D^k \times D^{m-k} \cong A_i,$$

where $D^k_{1/2}$ denotes the closed k–disc of radius $1/2$, see Figure 3.3. One says that M_i is obtained from M_{i-1} by 'attaching a k-handle' (or handle of index k). The description of a manifold via successive attaching of handles along the boundary and subsequent smoothing of corners is called a *handle presentation*, the manifold (with boundary) itself a *handlebody*.

26 3. THE h-PRINCIPLE FOR OPEN, INVARIANT RELATIONS

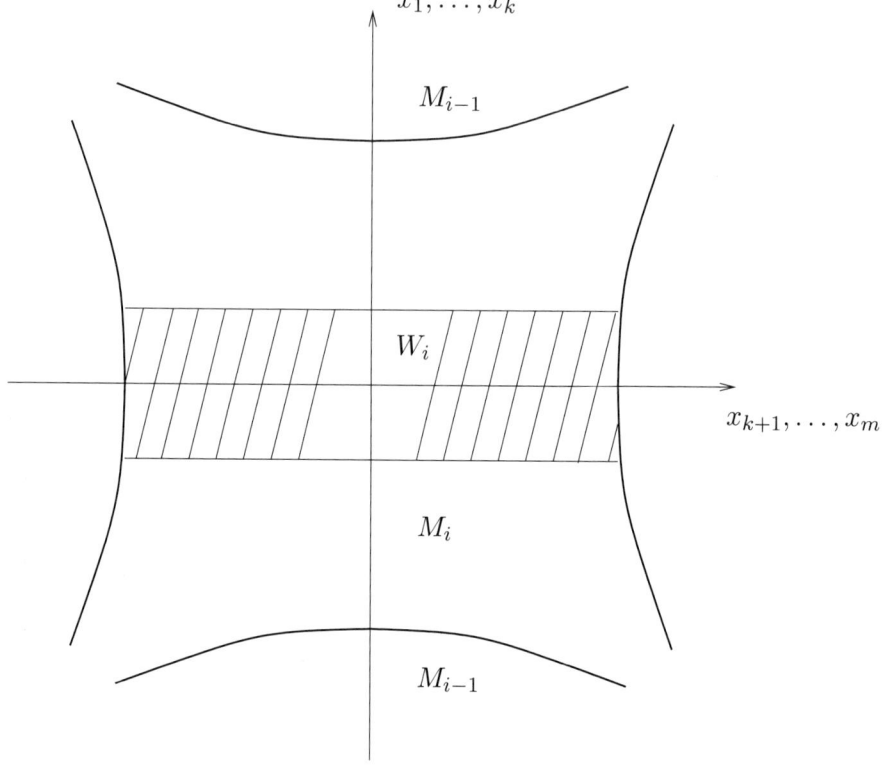

FIGURE 3.2. Neighbourhood of a critical point.

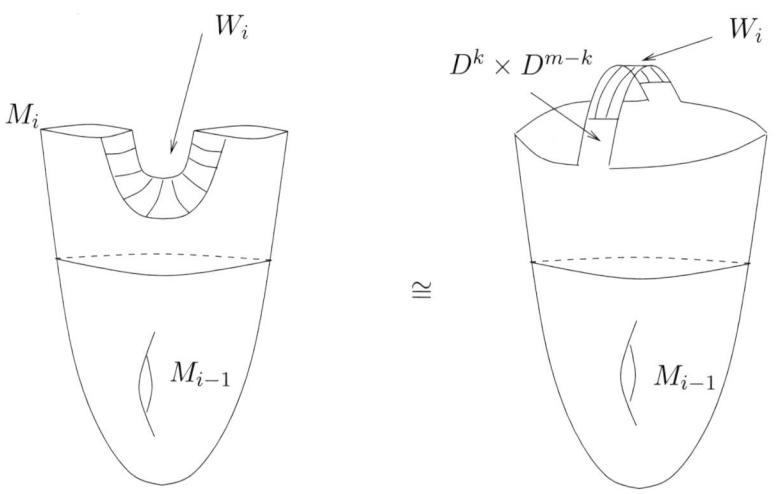

FIGURE 3.3. Attaching a handle.

3.4. PROOF OF THE THEOREM

Our aim now will be to prove Theorem 3.3 first for the manifold $M_1 \cong D^m$, and then to extend the result step by step over the exhaustion of M given by

$$D^m \cong M_1 \subset \ldots \subset M_{i-1} \subset M'_{i-1} \subset M_i \subset M'_i \subset \ldots \subset M.$$

(Clearly a little care will be necessary if the Morse function has infinitely many critical points.) Thus, the first step of the construction is the following proposition.

PROPOSITION 3.12. *Theorem 3.3 is true for the m–disc, i.e.*

$$j^r \colon \Gamma_0(E|M_1) \longrightarrow \Gamma(\mathcal{R}|M_1)$$

is a w.h.e.

The inductive procedure then hinges on the following two propositions, and a general lemma about Serre fibrations.

PROPOSITION 3.13. *The restriction maps*

$$\rho_0 \colon \Gamma_0(E|M'_{i-1}) \longrightarrow \Gamma_0(E|M_{i-1})$$

and

$$\rho \colon \Gamma(\mathcal{R}|M'_{i-1}) \longrightarrow \Gamma(\mathcal{R}|M_{i-1})$$

are w.h.e.'s and Serre fibrations.

Not surprisingly, this proposition holds more generally for restriction maps from any kind of collar neighbourhood, also with respect to a subset of the coordinates. Another particularly relevant case for us is that of a collarlike neighbourhood of $A_i - W_i$ in A_i. In the sequel, whenever we write a space of sections over a closed set of this kind, it will be understood that we mean sections defined over a collarlike neighbourhood (which, in certain situations, may depend on the section).

PROPOSITION 3.14. *The restriction maps*

$$\rho_0 \colon \Gamma_0(E|A_i) \longrightarrow \Gamma_0(E|A_i - W_i)$$

and

$$\rho \colon \Gamma(\mathcal{R}|A_i) \longrightarrow \Gamma(\mathcal{R}|A_i - W_i)$$

are Serre fibrations.

Here we only need to work with the explicit models for A_i and W_i described above. The part of this proposition about ρ_0 is what amounts to 'flexibility' in a more technical sense. In fact, Gromov [**15**] shows that it is possible to work with the weaker notion of 'microflexibility', where these restriction maps are required to be 'microfibrations', meaning that they have the homotopy lifting property for compact polyhedra not over the whole homotopy interval $[0,1]$, but only over some small interval $[0,\varepsilon]$. Over submanifolds which are 'sharply movable' by local diffeomorphisms of M that lift to $\mathcal{R}$-preserving diffeomorphisms of E^r, one then has flexibility. Thus, this allows to treat more general relations than invariant ones (where all local diffeomorphisms of M lift). The proof of Proposition 3.14 will give the reader some idea what should be the definition of 'sharply movable' and where the concept of 'microflexibility' enters into the argument.

LEMMA 3.15. *Consider the following commutative diagram,*

$$\begin{array}{ccc} X & \xrightarrow{g} & X' \\ \pi \downarrow & & \downarrow \pi' \\ B & \xrightarrow{\bar{g}} & B', \end{array}$$

where π and π' are Serre fibrations. Write

$$g_b = g|X_b \colon X_b \longrightarrow X'_{\bar{g}(b)}$$

for the restriction of g to a fibre X_b. If any two of g, $\bar{g}$ and g_b (for all/any $b \in B$) are w.h.e.'s, then so is the third.

PROOF. This follows immediately from the homotopy exact sequence for Serre fibrations and the Five Lemma. □

I defer the proof of Propositions 3.12 and 3.14 to the next section. The idea for the proof of Proposition 3.13 is rather simple, and Haefliger [**16**] quite sensibly leaves out details (and so does M. Adachi [**1**], who copies Haefliger's proof almost word for word). Still, some readers may wish to see these details, although these details should probably be skipped on a first reading (or better still: worked out by the reader), lest they obscure the beauty of the argument. Similar reasoning as in the proof of Proposition 3.13 is necessary in other parts of the proofs

3.4. PROOF OF THEOREM

of Propositions 3.12 and 3.14, as well as the remaining steps in proving Theorem 3.3. So the proof of Proposition 3.13 may be seen as a warm-up for these subsequent arguments.

PROOF OF PROPOSITION 3.13. To simplify notation we suppress the subscript $i-1$. The essential observation is that sections in $\Gamma \mathcal{R}|M$ and $\Gamma_0 E|M$, respectively, can be extended to a neighbourhood U of M in M', and that there is an isotopy, fixed on a smaller neighbourhood of M, deforming the identity of M' into an embedding of M' in U.

It is pretty clear that any argument that works for $\Gamma_0 E$ will work *a fortiori* for $\Gamma \mathcal{R}$, so I only present the former.

(i) surjectivity on π_k: Let an element of $\pi_k(\Gamma_0(E|M))$ be given, represented by a continuous map $S^k \to \Gamma_0(E|M)$. Interpret this as a section of the bundle E pulled back to $M \times S^k$ (under the projection $M \times S^k \to M$), of class C^r with respect to the M–factor, and whose r–jet with respect to the M–factor satisfies the relation $\mathcal{R}$ and is continuous with respect to the S^k–factor. I shall call such sections *admissible*.

Since $\mathcal{R}$ is open, this extends to an admissible section of E over a neighbourhood of $M \times S^k$ in $M' \times S^k$ (simply construct any extended section satisfying the differentiability requirements – either by *ad hoc* arguments or by invoking an appropriate version of Whitney's extension theorem [**31**] – and then choose the neighbourhood small enough so that the differential relation is still satisfied). Choose an embedding of $M' \times S^k$ into this neighbourhood, fixing the S^k–coordinate and equal to the identity in some smaller neighbourhood of $M \times S^k$. The invariance of $\mathcal{R}$ then allows to pull back the given (extended) section to an admissible section over $M' \times S^k$ whose restriction to $M \times S^k$ is the one we started with.

(ii) injectivity on π_k: Assume we have a continuous map $S^k \to \Gamma_0(E|M')$ that on restriction to M represents the trivial element of $\pi_k(\Gamma_0(E|M))$ (or the component containing a chosen base point in the case $k = 0$). This means that we have an admissible section of E over $M' \times S^k \cup M \times D^{k+1}$. Again we extend this to an admissible section in a neighbourhood. There is an embedding of $M' \times D^{k+1}$ into this neighbourhood, fixing the D^{k+1}–coordinate, and equal to the identity map in a smaller neighbourhood of $M' \times S^k \cup M \times D^{k+1}$ inside $M' \times D^{k+1}$ (think of M' as $M \cup \partial M \times [0, \delta]$ and observe that $D^{k+1} \times [0, \delta]$ embeds into a neighbourhood of $D^{k+1} \times \{0\} \cup S^k \times [0, \delta]$, leaving the D^{k+1}–coordinate fixed). Arguing as in (i), we see that the given map $S^k \to \Gamma_0(E|M')$ extends over D^{k+1}.

(iii) Serre fibration property: We have to consider the following commutative diagram, where Q is any compact polyhedron, $I = [0, 1]$, the horizontal maps are continuous, and $Q \times \{0\} \to Q \times I$ is the obvious inclusion:

$$\begin{array}{ccc} Q \times \{0\} & \longrightarrow & \Gamma_0(E|M') \\ \downarrow & & \downarrow \rho_0 \\ Q \times I & \longrightarrow & \Gamma_0(E|M). \end{array}$$

Our task is to show that there is a map $Q \times I \to \Gamma_0(E|M')$ making the resulting diagram commutative. Pull back E to $M' \times Q \times I$ under the projection onto the M'–factor. We define admissible sections of this bundle over subsets of $M' \times Q \times I$ in obvious analogy with (i).

The diagram above says that we have an admissible section over

$$M' \times Q \times \{0\} \cup M \times Q \times I.$$

Again thinking of M' as $M \cup \partial M \times [0, \delta]$, we see that there is an embedding of $M' \times Q \times I$ into a neighbourhood of the described set, keeping the Q– and I–coordinate fixed. The proof now concludes as in (i) and (ii).

This completes the proof of Proposition 3.13. $\square$

Taking Propositions 3.12 and 3.14 for granted, we can now proceed to prove Gromov's h-principle for open, invariant relations.

PROOF OF THEOREM 3.3. The proof is by stepwise extension over the exhaustion of M described earlier, and induction on the index of the relevant critical point of f.

The theorem is true for one 0–handle (i.e. an m–disc) by Proposition 3.12. Now assume that the theorem has already been proved for M_{i-1} and any handlebody containing only handles of index less than k, where k is the index of the handle we need to attach to M_{i-1} in order

3.4. PROOF OF THEOREM

to obtain M_i, that is,

$$M_i = M'_{i-1} \cup A_i, \ M'_{i-1} \cap A_i = A_i - W_i, \ A_i = D^k \times D^{m-k}.$$

Step 1: Consider the commutative diagram

$$\begin{array}{ccc} \Gamma_0(E|A_i) & \xrightarrow{j^r} & \Gamma(\mathcal{R}|A_i) \\ {\scriptstyle \rho_0}\downarrow & & \downarrow{\scriptstyle \rho} \\ \Gamma_0(E|A_i - W_i) & \xrightarrow{j^r} & \Gamma(\mathcal{R}|A_i - W_i). \end{array}$$

The interior of A_i is diffeomorphic to an open m–disc, so with Proposition 3.12 (and Proposition 3.13; cf. the remark after that proposition) we may conclude that the j^r at the top is a w.h.e.

According to Proposition 3.14, both ρ_0 and ρ are Serre fibrations.

We claim that the j^r at the bottom is a w.h.e. by the inductive assumption. Indeed, we have

$$\begin{aligned} A_i - W_i &= \{x \in D^k \colon 1/2 \leq |x| \leq 1\} \times D^{m-k} \\ &\cong S^{k-1} \times D^1 \times D^{m-k} \\ &\cong \left(D^{k-1} \times D^1 \cup_{S^{k-2} \times D^1} D^{k-1} \times D^1\right) \times D^{m-k} \\ &\cong D^m \cup_{S^{k-2} \times D^{m-k+1}} D^{k-1} \times D^{m-k+1} \\ &= 0\text{–handle} \cup (k-1)\text{–handle}, \end{aligned}$$

so the inductive assumption applies (again, Proposition 3.13 is implicitly invoked here).

From Lemma 3.15 we now conclude that j^r is a w.h.e. from the fibre of ρ_0 to that of ρ in the diagram above.

Step 2: Consider the commutative diagram

$$\begin{array}{ccc} \Gamma_0(E|M_i) & \xrightarrow{j^r} & \Gamma(\mathcal{R}|M_i) \\ {\scriptstyle \rho_0}\downarrow & & \downarrow{\scriptstyle \rho} \\ \Gamma_0(E|M'_{i-1}) & \xrightarrow{j^r} & \Gamma(\mathcal{R}|M'_{i-1}). \end{array}$$

This is the pull-back diagram of the one considered in Step 1 under the restriction to A_i (observe that $M'_{i-1} \cap A_i = A_i - W_i$). Hence the ρ_0, ρ in that last diagram are also Serre fibrations, and j^r is a w.h.e. on the fibre. Moreover, the j^r at the bottom (i.e. on the level of M'_{i-1}) is a w.h.e. by the inductive assumption and Proposition 3.13.

Invoking Lemma 3.15 once again, we conclude that j^r is also a w.h.e. on the level of M_i.

This completes the proof of Theorem 3.3 provided f has only finitely many critical points, i.e. if we have a handle presentation of M with only finitely many handles.

For the case that we have an infinite number of handles, we need to recall the concept of an inverse limit. Assume we are given a sequence $(B_i)_{i=0}^{\infty}$ of topological spaces, and continuous maps $r_{ik}\colon B_k \to B_i$ whenever $k \geq i$, such that $r_{ik} = r_{ij} \circ r_{jk}$ for $k \geq j \geq i$. Then the *inverse* (or *projective*) *limit* B of the system (B_i, r_{ik}) is the subset $\varprojlim(B_i, r_{ik})$ of the cartesian product $\prod_{i=0}^{\infty} B_i$ consisting of the points $(\ldots, b_2, b_1, b_0)$ such that $b_i = r_{ik}(b_k)$ for $k \geq i$, topologised as a subset of the product space. There are canonical maps $r_i\colon B \to B_i$.

Assume we have a second system (B_i', r_{ik}') of the same kind, and continuous maps $d_i : B_i \to B_i'$ giving rise to an infinite commutative diagram, i.e. with $d_i \circ r_{ik} = r_{ik}' \circ d_k$. Then the *inverse limit* of the d_i is the unique continuous map $d\colon B \to B'$ defined by $r_i' \circ d = d_i \circ r_i$.

LEMMA 3.16. *If the r_{ik} and r_{ik}' are Serre fibrations and the d_i w.h.e.'s, then d is a w.h.e.*

For a proof see [**26**, App. I, Lemma 2]. Now set $B_i = \Gamma_0(E|M_i)$ and $B_i' = \Gamma(\mathcal{R}|M_i)$, and d_i equal to the restriction of j^r to the appropriate space. Then B is identified as $\Gamma_0 E$ and B' as $\Gamma\mathcal{R}$, and the inverse limit of the j^r is of course the map $j^r\colon \Gamma_0 E \to \Gamma\mathcal{R}$, and thus a w.h.e. by the lemma just stated.

This concludes the proof of Theorem 3.3. □

3.5. Further details of the proof

In this section we provide the proofs of Propositions 3.12 and 3.14.

PROOF OF PROPOSITION 3.12. In order to simplify notation we assume from the start that E is a bundle over D^m, hence a trivial bundle $E = D^m \times F$. By assumption, $\mathcal{R}$ is a subbundle of E, and hence, as a bundle over D^m, also trivial. Write $F_{\mathcal{R}}$ for the fibre of $\mathcal{R}$.

We thus have the identifications $\Gamma^r(E) = C^r(D^m, F)$ and, in the same vein, $E_x^r = J_x^r(D^m, F)$, the space of r–jets of germs at $x \in D^m$ of maps $D^m \to F$. Similarly, $\Gamma\mathcal{R} = C^0(D^m, F_\mathcal{R})$.

(i) We observe that the restriction map

$$\rho\colon \Gamma\mathcal{R} \to \Gamma(\mathcal{R}|0) \equiv F_\mathcal{R}$$

is a w.h.e. This can be seen quite explicitly. Surjectivity on π_k follows from the existence of constant sections: Given a map $f_0\colon S^k \to F_\mathcal{R}$, it can be extended to a map $f\colon S^k \to \Gamma\mathcal{R}$ with $\rho(f) = f_0$ simply by setting $f(s)(x) = f_0(s)$ for all $s \in S^k$ and $x \in D^m$. Injectivity on π_k is proved as follows: Assume we have $f \in \pi_k(\Gamma\mathcal{R})$ such that $f(.)(0)\colon S^k \to F_\mathcal{R}$ is homotopically trivial, i.e. such that $f(s)(0)$ can be defined for all $s \in D^{k+1}$. Then define $\widetilde{f}\colon D^{k+1} \times D^m \to F_\mathcal{R}$ by

$$\widetilde{f}(s,x) = \begin{cases} f(s/|s|, (2|s|-1)x) & \text{for } 1/2 \leq |s| \leq 1, \\ f(2s, 0) & \text{for } 0 \leq |s| \leq 1/2. \end{cases}$$

This shows that f is homotopically trivial.

(ii) The considerations in (i) imply that it is sufficient to prove that

$$j_0^r = \rho \circ j^r\colon \Gamma_0 E \longrightarrow F_\mathcal{R}$$

is a w.h.e.

(ii, 1) surjectivity on π_k: Fix an embedding $F \hookrightarrow \mathbb{R}^q$ for suitably large q, and identify F with its image in $\mathbb{R}^q$ under this embedding. Let $\nu F \subset \mathbb{R}^q$ be a tubular neighbourhood of F in $\mathbb{R}^q$, and write $\pi\colon \nu F \to F$ for the projection along the fibres of this tubular neighbourhood.

Given an element of $\pi_k(F_\mathcal{R})$, represent it by a map

$$f\colon S^k \to F_\mathcal{R} \subset J_0^r(D^m, F) \subset J_0^r(D^m, \mathbb{R}^q).$$

Let $\widetilde{w}_s\colon D^m \to \mathbb{R}^q$ be the polynomial representative of $f(s)$, that is, the unique polynomial of degree r with $j_0^r \widetilde{w}_s = f(s)$. Our aim is to construct an r times continuously differentiable map $D^m \to F$ representing the same r–jet at 0.

Let $U \subset D^m$ be a disc neighbourhood of 0 such that $\widetilde{w}_s(U) \subset \nu F$ and such that the map w_s defined by $w_s = \pi \circ \widetilde{w}_s|U$ is an element of $\Gamma_0(E|U)$. This is possible by the openness of $\mathcal{R}$ and the compactness of the parameter space S^k. Observe that $\widetilde{w}_s$ is tangent to F of order r at $x = 0$, hence $j_0^r w_s = j_0^r \widetilde{w}_s$.

Let $h\colon D^m \to U$ be a diffeomorphism that equals the identity map near $0 \in D^m$. Recall the commutative diagram for the continuous extension Φ (which exists by the invariance of $\mathcal{R}$):

$$\begin{array}{ccc} E & \xrightarrow{\Phi(h)} & E|U \\ \downarrow & & \downarrow \\ D^m & \xrightarrow{h} & U. \end{array}$$

Now set $g_s = \Phi(h)^{-1} \circ w_s \circ h$. Then, for any $x \in D^m$,

$$\begin{aligned} j_x^r g_s &= j_x^r(\Phi(h)^{-1} \circ w_s \circ h) \\ &= \Phi^r(h^{-1})(j_{h(x)}^r w_s). \end{aligned}$$

Since $w_s \in \Gamma_0(E|U)$, we have $j_{h(x)}^r w_s \in \mathcal{R}_{h(x)}$, whence $j_x^r g_s \in \mathcal{R}_x$ by the invariance of $\mathcal{R}$. This means $g_s \in \Gamma_0 E$. Furthermore, by construction of h and w we have

$$j_0^r g_s = \Phi^r(h^{-1})(j_0^r w_s) = j_0^r w_s = j_0^r \widetilde{w}_s = f(s).$$

(ii, 2) injectivity on π_k: This is proved by similar arguments; here is a brief sketch. Assume we are given a parametric family $g_s \in \Gamma_0 E$, where $s \in S^k$, again regarded as a family of C^r–maps $g_s\colon D^m \to F$, with the property that the family of r–jets, $f_s = j_0^r g_s \in F_\mathcal{R}$, extends to a family defined for all $s \in D^{k+1}$. The construction in (ii) yields a family $\overline{g}_s\colon D^m \to F$, defined for $s \in D^{k+1}$, with the property that for $s \in S^k$ we have $j_0^r \overline{g}_s = j_0^r g_s = f_s$. Apply the construction as in (ii) to the linear interpolation between $\overline{g}_s$ and g_s, for $s \in S^k$, and regard the homotopy parameter as the radial coordinate in an annulus $S^k \times [0, 1]$. Gluing this annulus with D^{k+1} defines the extension of the $g_s \in \Gamma_0 E$ to a family defined for all s in a $(k+1)$–disc $D^{k+1} \cup S^k \times [0, 1]$. $\square$

Before we turn to the proof of Proposition 3.14, here is a simple example where 'flexibility' in the sense of that proposition fails, namely, immersions of a 1–dimensional manifold in the real line. Let $A = D^1$ and $W = \mathrm{Int}(D_{1/2}^1)$, and define, with $t \in [0, 1]$, a regular homotopy of

immersions
$$f_t\colon A - W \longrightarrow \mathbb{R}$$
$$x \longmapsto \begin{cases} x, & x > 0, \\ x + 2t, & x < 0. \end{cases}$$

Then f_0 lifts to an immersion $A \to \mathbb{R}$ (the obvious inclusion), but f_1 does not. So $\rho_0\colon \mathcal{I}mm(A,\mathbb{R}) \to \mathcal{I}mm(A-W,\mathbb{R})$ is not a Serre fibration.

PROOF OF PROPOSITION 3.14. We first introduce some notation:
$$\begin{aligned} D_a^k &= \{x \in \mathbb{R}^k \colon |x| \le a\}, \\ D_{[a,b]}^k &= \{x \in \mathbb{R}^k \colon a \le |x| \le b\}, \\ S_a^{k-1} &= \{x \in \mathbb{R}^k \colon |x| = a\}. \end{aligned}$$

Furthermore, we make the following identifications, where k is some natural number smaller than m, fixed throughout this proof:
$$\begin{aligned} A &= D_2^k \times D^{m-k}, \\ A - W &= D_{[1,2]}^k \times D^{m-k}, \\ E|A &= A \times F, \\ \mathcal{R}|A &= A \times F_\mathcal{R}. \end{aligned}$$

As in the preceding proof, we think of sections of E or $\mathcal{R}$ over A as maps from A into the relevant fibre.

(i) The restriction map $\rho\colon \Gamma(\mathcal{R}|A) \to \Gamma(\mathcal{R}|A-W)$ is a Serre fibration: To show that, we have to start from a commutative diagram

$$\begin{array}{ccc} Q \times \{0\} & \longrightarrow & \Gamma(\mathcal{R}|A) \\ \downarrow & & \downarrow \\ Q \times I & \longrightarrow & \Gamma(\mathcal{R}|A - W), \end{array}$$

where Q denotes again a compact polyhedron, and we want to find a map $Q \times I \to \Gamma(\mathcal{R}|A)$ making the diagram commutative. We may interpret the given diagram as saying that $f_{q,y}(rs,t) \in F_\mathcal{R}$ is defined, depending continuously on $q \in Q$, $y \in D^{m-k}$, $s \in S^{k-1}$, and
$$\begin{cases} r \in [0,2] \text{ for } t = 0, \\ r \in [1,2] \text{ for } t \in I. \end{cases}$$

Since
$$D_2^k \times \{0\} \cup D_{[1,2]}^k \times I$$
is a deformation retract of $D_2^k \times I$, we can define $f_{q,y}(x,t)$ (as a continuous map in all variables) for all $x \in D_2^k$ and $t \in I$, extending the given map.

(ii) The restriction map $\rho_0 \colon \Gamma_0(E|A) \to \Gamma_0(E|A-W)$ is a Serre fibration: In order to deal with this problem, we introduce the concept of an admissible map, very similar to that in the proof of Proposition 3.13: Let $\widetilde{Q}$ be some parameter space (in practice this will be a polyhedron Q or a polyhedron times the homotopy interval I), and let V be a subset of $\mathbb{R}^m$. Then a map $V \times \widetilde{Q} \to F$ will be called *admissible* if it is of class C^r for any fixed parameter, and the r–jet with respect to the V–factor is continuous in the parameter and satisfies the relation $\mathcal{R}$.

Thus, our problem can be reformulated as follows: Given are admissible maps
$$f \colon D_{[1,2]}^k \times D^{m-k} \times Q \times [0,1] \to F$$
and
$$g_0 \colon D_2^k \times D^{m-k} \times Q \times \{0\} \to F,$$
with $f = g_0$ where both are defined. In the description of f, the radial coordinate[2] r in D^k is regarded as a collar parameter; and by our conventions introduced after Proposition 3.13 and the compactness of the parameter space $Q \times [0,1]$ it is understood that f is actually defined on a collarlike neighbourhood of
$$D_{[\alpha,2]}^k \times D^{m-k} \times Q \times [0,1]$$
for some $\alpha < 1$ (and $f = g_0$ where both are defined also for this larger domain of definition of f).[3] We want to construct an admissible extension
$$g \colon D_2^k \times D^{m-k} \times Q \times [0,1] \to F.$$

---

[2] There should be little grounds for confusing the two different meanings of 'r' in this proof.

[3] Without this convention about sections being defined in neighbourhoods of closed sets, step (a) in Haefliger's argument [16, p. 138] (the part about $f = g_0$ where both are defined – or f' in Haefliger's notation there) needs a little more care; again a suitable version of Whitney's extension theorem may be used.

Step 1: We claim that there is a sequence $0 = t_0 < t_1 < \ldots < t_n = 1$ and admissible maps μ_i, $0 \leq i \leq n-1$, defined on a neighbourhood of

$$D^k_{[\alpha,2]} \times D^{m-k} \times Q \times [t_i, t_{i+1}],$$

such that

$$\mu_i(x,y,q,t) = \begin{cases} f(x,y,q,t) & \text{for } t = t_i \text{ or } x \text{ in a nbhd. of } D^k_{[1,2]}, \\ f(x,y,q,t_i) & \text{for } x \text{ in a nbhd. of } S^{k-1}_\alpha. \end{cases}$$

Without the condition that the r–jet on slices $(q,t) = \text{const.}$ satisfy the relation $\mathcal{R}$ this is clearly possible by interpolating with a smooth bump function (in the radial parameter in D^k) between $f(x,y,q,t)$ and $f(x,y,q,t_i)$. By the openness of the relation $\mathcal{R}$ and the compactness of the homotopy interval (and the other parameter spaces), it is also possible within the class of admissible functions. Notice that in the time interval $[0, t_1]$ the homotopy extends to all $x \in D^k_2$ as a homotopy constant in t for $r < \alpha$. This is what earlier was referred to as 'microflexibility', which is thus clearly satisfied for open relations, cf. [15, p. 41]. As the next step will show, this microflexibility is really all that is needed to prove the h-principle for invariant relations on open manifolds.

Step 2: We now construct the desired map g by an inductive procedure over this sequence of subintervals of $[0, 1]$. It is the microflexibility of the relation that allows to start this inductive procedure, that is, we define

$$g_1(x,y,q,t) = \begin{cases} \mu_0(x,y,p,t) & \text{for } x \in D^k_{[\alpha,2]}, t \leq t_1, \\ g_0(x,y,q,0) & \text{for } x \in D^k_\alpha, t \leq t_1. \end{cases}$$

For the inductive step we need to show that we have a form of microflexibility over the interval $[t_i, t_{i+1}]$. The idea is that the existence of the μ_i (some weaker form of microflexibility, because μ_i is not defined for all $x \in D^k_2$ even at $t = t_i$) suffices, provided we are dealing with handles of index $k < m$, which leaves us 'room to move around'. In the more general context of Gromov's book [15], this gives rise to the notion 'sharply movable'.

Thus, we assume that an admissible

$$g_i \colon D^k_2 \times D^{m-k} \times Q \times [0, t_i] \to F$$

has been constructed with $g_i = f$ in a neighbourhood of

$$D^k_{[\beta,2]} \times D^{m-k} \times Q \times [0, t_i]$$

for some β with $\alpha < \beta < 1$.

Let $U \subset D^k_2 \times D^{m-k}$ be a neighbourhood of $D^k_{[\alpha,\beta]} \times \{0\}$ on which both μ_i and f are defined for all $q \in Q$ and $t \in [t_i, t_{i+1}]$, see Figure 3.4.

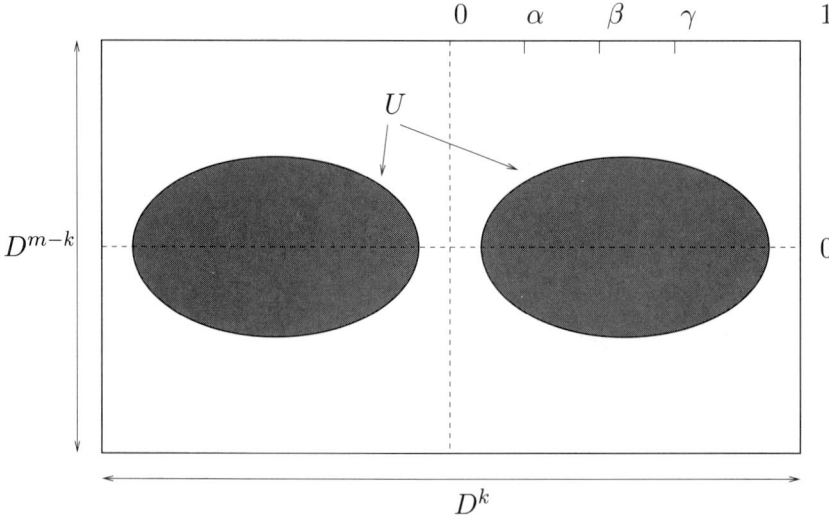

FIGURE 3.4. The neighbourhood U.

Now we come to the point where the condition $k < m$ enters crucially into the argument. Namely, we can find an isotopy ψ_t of $D^k_2 \times D^{m-k}$, for $0 \leq t \leq t_i$, satisfying the following conditions, see Figure 3.5:

$$\begin{aligned}\psi_t &= \text{id} \quad \text{outside } U, \\ &\quad \text{on a nbhd. of } S^{k-1}_\beta \times \{0\}, \\ &\quad \text{on a nbhd. of } S^{k-1}_1 \times \{0\}, \\ &\quad \text{for } t \leq t_i/2,\end{aligned}$$

and furthermore

$$\psi_{t_i}(S^{k-1}_\gamma \times \{0\}) = S^{k-1}_\alpha \times \{0\} \text{ for some } \beta < \gamma < 1.$$

Now define $\bar{g}_{i+1}$ on a small neighbourhood V of

$$(D^k_2 \times \{0\}) \cup (D^k_{[1,2]} \times D^{m-k}) \times Q \times [0, t_{i+1}]$$

3.5. FURTHER DETAILS OF THE PROOF

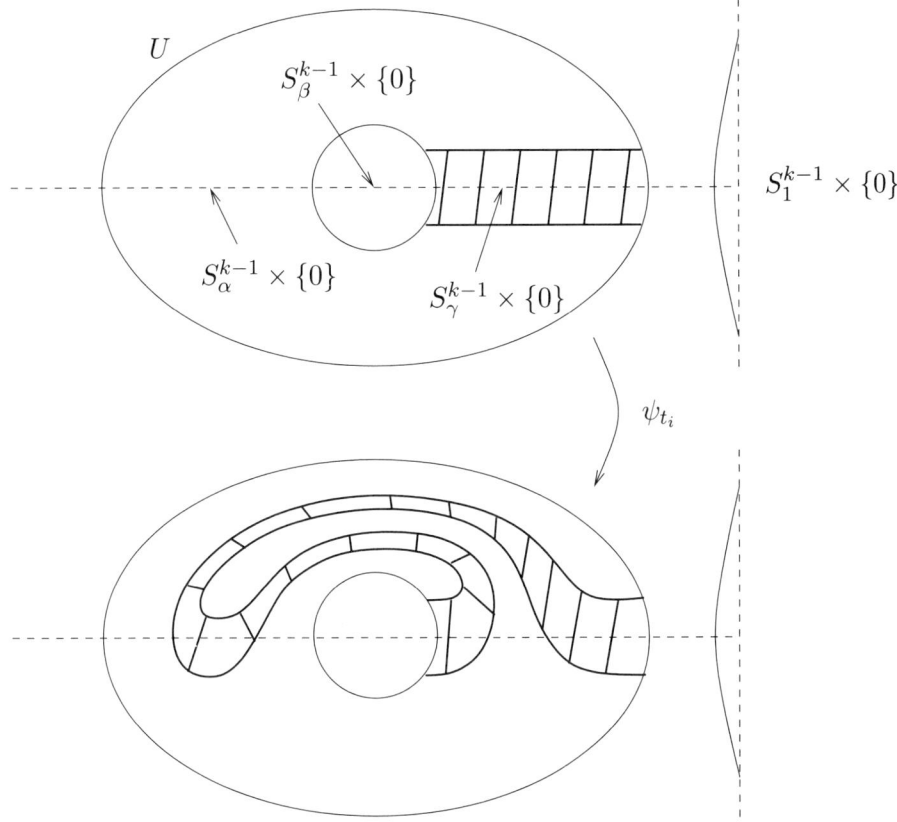

FIGURE 3.5. The isotopy ψ_{t_i}.

(see Figure 3.6) and depending on (r, t) lying in one of the regions I, II, III, IV (see Figure 3.7) as follows:

$$
\begin{aligned}
\text{I}: \quad & \overline{g}_{i+1} = g_i \\
\text{II}: \quad & \overline{g}_{i+1}(x, y, q, t) = \Phi(\psi_t)^{-1} f(\psi_t(x, y), q, t) \\
\text{III}: \quad & \overline{g}_{i+1}(x, y, q, t) = \Phi(\psi_{t_i})^{-1} \mu_i(\psi_{t_i}(x, y), q, t) \\
\text{IV}: \quad & \overline{g}_{i+1}(x, y, q, t) = \overline{g}_{i+1}(x, y, q, t_i)
\end{aligned}
$$

It is a straightforward check that these maps coincide along the boundaries of the four domains of definition and define, thanks to the invariance of $\mathcal{R}$, an admissible map, provided V is chosen small enough:

- I/II: The isotopy ψ_t does not move points in a neighbourhood of the sphere $S_\beta^{k-1} \times \{0\}$, so we only need to ensure that the intersection $V \cap \{r = \beta\}$ is contained in this neighbourhood.

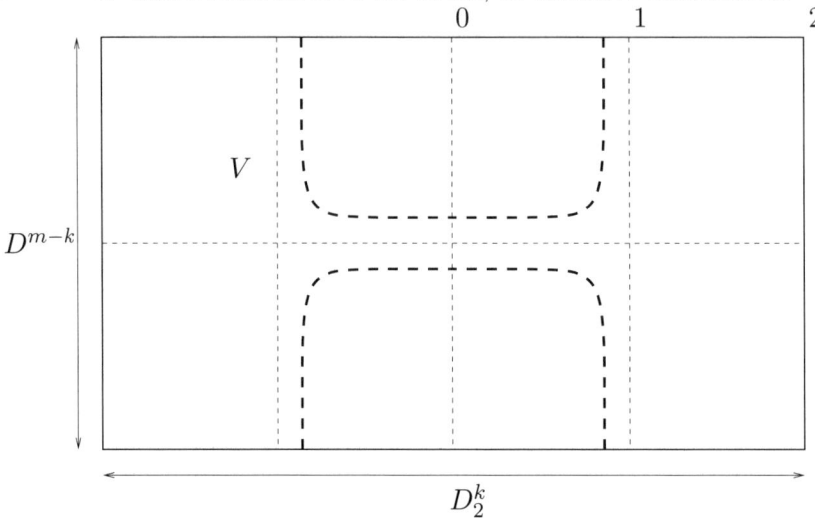

FIGURE 3.6. The neighbourhood V.

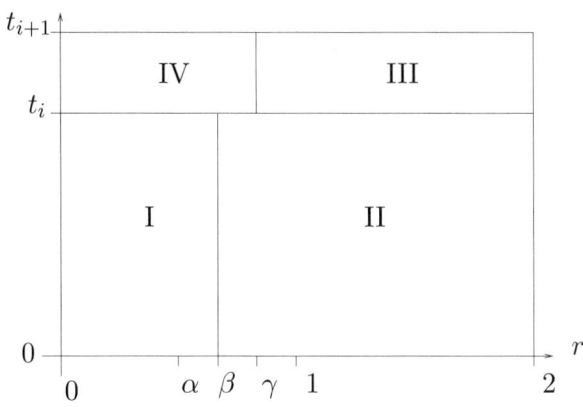

FIGURE 3.7. Definition of $\overline{g}_{i+1}$.

- II/III: Simply observe that $\mu_i = f$ (where both are defined) for $t = t_i$, and ψ_{t_i} only moves points in U, which was chosen exactly so that both μ_i and f both are defined.
- III/IV: Choose V small enough such that ψ_{t_i} moves a neighbourhood of $V \cap \{r = \gamma\}$ into the region where μ_i is independent of t.
- IV/I: Here the two definitions trivially match up continuously in the t–variable.

Notice that for $t \leq t_i/2$ (where ψ_t is the identity map), the map $\overline{g}_{i+1}$ is actually defined for all $(x, y) \in D_2^k \times D^{m-k}$.

3.5. FURTHER DETAILS OF THE PROOF

Finally, to extend the definition of $\bar{g}_{i+1}$ from points with $(x,y) \in V$ to an admissible map g_{i+1} defined for all $(x,y) \in D_2^k \times D^{m-k}$ we choose an isotopy of embeddings

$$\varphi_t \colon D_2^k \times D^{m-k} \hookrightarrow D_2^k \times D^{m-k}$$

such that

$$\varphi_0 = \mathrm{id},$$
$$\varphi_t = \mathrm{id} \text{ near } D_{[1,2]}^k \times D^{m-k},$$

and

$$\varphi_t(D_2^k \times D^{m-k}) \subset V \text{ for } t \geq t_i/2,$$

so that $D_2^k \times D^{m-k}$ has been moved into V before the isotopy ψ_t takes effect. Then, for all $(x,y,q,t) \in D_2^k \times D^{m-k} \times Q \times [0, t_{i+1}]$, the point $(\varphi_t(x,y), q, t)$ lies in a region where $\bar{g}_{i+1}$ has been defined, so we may set

$$g_{i+1}(x,y,q,t) = \Phi(\varphi_t)^{-1} \bar{g}_{i+1}(\varphi_t(x,y), q, t).$$

Again thanks to the invariance of $\mathcal{R}$ this defines an admissible map. We have thus completed the inductive step, and the proof of Proposition 3.14. □

CHAPTER 4

Convex Integration Theory

Most h-principles that can be proved by the method of convex integration can also be proved with the previously discussed covering homotopy method, sometimes at the expense of dealing with a slightly altered set-up. There are cases, however, where the advantage of convex integration lies in the fact that the dimension conditions are less restrictive, and that it yields better approximation (the covering homotopy method usually giving only C^0–approximations). For this reason convex integration theory has been employed successfully in proving h-principles for certain geometric structures on closed manifolds and embedding problems, in particular C^1–isometric ones.

Before introducing various technical terms and describing the main general h-principle that can be proved via convex integration, I outline the key idea behind this method. For an arbitrary subset $A \subset \mathbb{R}^n$, let $\mathrm{Conv}(A)$ denote the convex hull of A. If $f \in C^1(S^1, \mathbb{R}^n)$ with $f' \in C^0(S^1, A)$, then $0 \in \overline{\mathrm{Conv}(A)}$, since the integral $0 = \int_{S^1} f'(s)\,ds$ is a limit of Riemann sums

$$\sum_i f'(s_i)\,\Delta s_i \in \mathrm{Conv}(A).$$

Conversely, I claim that if A is path-connected and $\mathrm{Conv}(A)$ contains a neighbourhood of 0, then there is an $f \in C^1(S^1, \mathbb{R}^n)$ with $f' \in C^0(S^1, A)$. Later we shall see that there is in fact an h-principle for maps $f \in C^1(S^1, \mathbb{R}^n)$ satisfying the differential relation $f' \in C^0(S^1, A)$.

Indeed, the assumption on A implies that we can find a step function $\varphi \colon [0,1] \to A$ with $\int_0^1 \varphi(s)\,ds = 0$. As we shall see below in greater generality, the conditions that A be connected and $0 \in \mathrm{Int}\,\mathrm{Conv}(A)$ allow for φ to be adjusted so as to be continuous. Then the desired f is found by defining $f(t) = \int_0^t \varphi(s)\,ds$ for $t \in S^1 = \mathbb{R}/\mathbb{Z}$.

4.1. The h-principle for open, ample relations

We now turn to the general set-up. We shall formulate the *relative* h-principle, because in applications this is usually the most effective version. So we consider a smooth fibre bundle $p\colon E \to M$, a closed subset $W \subset M$, and a differential relation $\mathcal{R} \subset E^r$, which is not assumed to be a subbundle.

Let $\sigma_0 \in \Gamma_0(E|\widetilde{W})$, that is, $\sigma_0 \in \Gamma^r(E|\widetilde{W})$ with $j^r\sigma_0 \in \Gamma^0(\mathcal{R}|\widetilde{W})$, where $\widetilde{W}$ is some open neighbourhood of W. We then define the spaces

$$\Gamma(\mathcal{R}, \sigma_0, W) = \{\varphi \in \Gamma^0\mathcal{R} \colon \varphi = j^r\sigma_0 \text{ near } W\}$$

and

$$\Gamma_0(E, \sigma_0, W) = \{\sigma \in \Gamma^r E \colon j^r\sigma \in \Gamma^0\mathcal{R}, \sigma = \sigma_0 \text{ near } W\}.$$

DEFINITION 4.1. We say that $\mathcal{R}$ satisfies the **relative (parametric) h-principle** (or **h-principle for extensions**) if

$$j^r\colon \Gamma_0(E, \sigma_0, W) \longrightarrow \Gamma(\mathcal{R}, \sigma_0, W)$$

is a w.h.e. for all closed subsets $W \subset M$ and all σ_0.

DEFINITION 4.2. Let L be an affine space. A subset $A \subset L$ is **ample** if each pathwise connected component A_0 of A satisfies $\mathrm{Conv}(A_0) = L$. The empty set is regarded as ample.

It should be clear from the 'key idea' described earlier that a differential relation $\mathcal{R}$ whose fibres $\mathcal{R}_x$ are ample in E^r_x has a good chance of satisfying an h-principle: a section of E can be approximated by a section whose jet satisfies the relation $\mathcal{R}$, because we can realise elements in E^r_x by taking an average over elements in R_x. The full truth is a little more subtle, and the notation alone can become quite daunting. For this reason I restrict myself to the case of 1–jets.

Recall that we wrote p^1_0 for the natural projection $E^1 \to E$. Choose local coordinates $x^1, \ldots, x^m$ in M near some point $x_0 = (x_0^1, \ldots, x_0^m)$, and let $y^1, \ldots, y^q$ be local coordinates in the fibre $p^{-1}(x_0) = E_{x_0}$ near some point

$$\widetilde{x}_0 = (x_0^1, \ldots, x_0^m, y_0^1, \ldots, y_0^q).$$

Then $(p_0^1)^{-1}(\widetilde{x}_0)$ has affine coordinates (a_{ki}), $1 \leq i \leq m$, $1 \leq k \leq q$, corresponding to the 1–jet represented by

$$y^k = y_0^k + \sum_{i=1}^m a_{ki}(x^i - x_0^i), \quad k = 1, \ldots, q.$$

Given a choice of coordinate direction x^i, we can define intermediate 'perp' jets as equivalence classes of sections that coincide along the slices $x^i = \text{const.}$, i.e. we define

$$j_{x_0}^\perp \sigma_1 = j_{x_0}^\perp \sigma_2 \;:\Longleftrightarrow\; \sigma_1|_{x^i=x_0^i} = \sigma_2|_{x^i=x_0^i}.$$

Equivalently, two local sections define the same perp jet corresponding to the coordinate direction x^i if their partial derivatives in the other coordinate directions coincide, that is, $a_{kj}^1 = a_{kj}^2$ for $j \neq i$.

These perp jets define an intermediate bundle $E^\perp = E^{\perp i}$, and intermediate fibrations

$$E^1 \xrightarrow{p_\perp^1} E^\perp \xrightarrow{p_0^\perp} E,$$

where $p_\perp^1(j^1\sigma) = j^\perp \sigma$. Observe that the fibres of $p_\perp^1$ are q–dimensional affine spaces. So we obtain a decomposition of the fibres of p_0^1, which are affine spaces of dimension mq, into a disjoint union of parallel q–dimensional affine subspaces. This decomposition is independent of the choice of fibre coordinates $y^1, \ldots, y^q$.

DEFINITION 4.3. The fibres of $p_\perp^1$ are called **principal subspaces** in E^1 (for the coordinate x^i).

The relation $\mathcal{R}$ is **ample in the coordinate directions** if the intersection of $\mathcal{R}$ with any principal subspace, for any of the n coordinates $x^1, \ldots, x^m$, is ample in this subspace.

THEOREM 4.4 (Gromov). *Suppose $\mathcal{R} \subset E^1$ is open. If, for every $\widetilde{x} \in E$, there are local coordinates about $p(\widetilde{x})$ such that $\mathcal{R}$ is ample in the coordinate directions for all the fibres of p_0^1 close to that over $\widetilde{x}$, then $\mathcal{R}$ satisfies the relative parametric h-principle.*

Here is a simple example illustrating this theorem. Let $M = \mathbb{R}$ and E be the trivial bundle $\mathbb{R} \times \mathbb{R}^3$. Then E^1 can be naturally identified with $\mathbb{R} \times \mathbb{R}^3 \times \mathbb{R}^3$. Because of $\dim M = 1$ we have only one coordinate direction to choose from, and in fact $E^\perp = E$. Given any $\varepsilon > 0$, define a differential relation $\mathcal{R}$ by

$$\mathcal{R} = \{(t, y, y') \in E^1 \colon |(y_1')^2 + (y_2')^2 - (y_3')^2| < \varepsilon\}.$$

This relation satisfies the assumptions of the preceding theorem, so we conclude that any continuous curve γ_0 in $\mathbb{R}^3$ is homotopic to a C^1–curve γ_1 whose velocity lies inside the thickened cone

$$\{|(y_1')^2 + (y_2')^2 - (y_3')^2| < \varepsilon\}.$$

In fact, a stronger version of Gromov's theorem holds, and this is easily seen to be true in this example (see Figure 4.1): The curve γ_1 may be chosen C^0–close to γ_0.

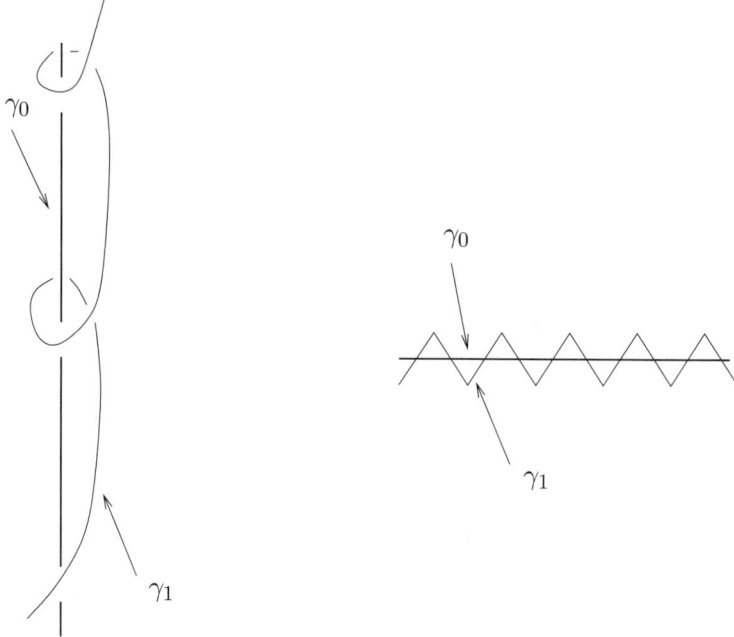

FIGURE 4.1. An illustration of convex integration.

An instructive example concerning the 1–parametric version of Theorem 4.4 is the classification of C^1–immersions $S^1 \to \mathbb{R}^2$ up to regular homotopy. Let $M = S^1 = \mathbb{R}/2\pi\mathbb{Z}$ and E be the trivial bundle $S^1 \times \mathbb{R}^2$, so that sections of E correspond to maps $S^1 \to \mathbb{R}^2$. The 1–jet bundle E^1 will be identified with $S^1 \times \mathbb{R}^2 \times \mathbb{R}^2$. As in the previous example we have $E^\perp = E$ since $\dim M = 1$.

Given $f \in \mathcal{I}mm(S^1, \mathbb{R}^2)$, its tangential map $Tf \in \mathcal{M}on(TS^1, T\mathbb{R}^2)$ can be thought of, up to homotopy, as a map
$$S^1 \longrightarrow \mathbb{R}^2 - \{0\} \simeq S^1.$$

The following result of Whitney [32], cf. Proposition 3.4, says that regular homotopy classes of immersions $S^1 \to \mathbb{R}^2$ are classified by the degree of their tangential map.

PROPOSITION 4.5 (Whitney). *Let $f_0, f_1 \colon S^1 \to \mathbb{R}^2$ be immersions and $n_0, n_1 \in \mathbb{Z}$ the degree of their respective tangential map. Then f_0 is regularly homotopic to f_1 if and only if $n_0 = n_1$.*

The 'only if' part is obvious, since regularly homotopic immersions have homotopic tangential maps. In the language of differential relations, the 'if' part says that the 1–jet map
$$\mathcal{I}mm(S^1, \mathbb{R}^2) \longrightarrow \mathcal{M}on(TS^1, T\mathbb{R}^2)$$

is injective on π_0. Rephrased like that, it is recognised as an immediate consequence of Theorem 4.4: In the given situation, the principal subspaces are 2–dimensional vector spaces, and the intersection of the immersion relation with these vector spaces equals $\mathbb{R}^2 - \{0\}$, which is ample in $\mathbb{R}^2$.

Here is the convex integration idea for giving a direct proof of Whitney's theorem. Assume that $n_0 = n_1$. By a regular homotopy of f_0 and f_1 we may assume further that

- $f_0(0) = f_1(0) = 0$.
- Both f_0 and f_1 have total length 2π and are parametrised by arc length.

The first point is achieved by a translation of $\mathbb{R}^2$, the second by a homothety of $\mathbb{R}^2$ and reparametrising the curves.

The assumption $n_0 = n_1$ implies that there is a homotopy
$$h_\tau \colon \mathbb{R}/2\pi\mathbb{Z} = S^1 \longrightarrow S^1 \subset \mathbb{R}^2, \ \tau \in [0,1],$$

with $h_0 = f_0'$ and $h_1 = f_1'$. Set
$$\widetilde{f}_\tau(t) = \int_0^t h_\tau(s) \, ds.$$

Then $\tilde{f}_\tau = f_\tau$ for $\tau = 0, 1$, but for $\tau \neq 0, 1$ the curve $\tilde{f}_\tau$ will not, in general, be closed. This is where we need convex integration. Consider the closed curve $\tau \mapsto \tilde{f}_\tau(1)$ in $\mathbb{R}^2$. Since $\mathbb{R}^2 - \{0\}$ is open and ample in $\mathbb{R}^2$, one can find a family χ_τ of functions, defined on an interval and taking values in $\mathbb{R}^2 - \{0\}$, with initial value $h_\tau(2\pi)$ and total integral equal to

$$-\int_0^{2\pi} h_\tau(s)\, ds.$$

The concatenation of h_τ and χ_τ, suitably parametrised, will then integrate to a regular homotopy between f_0 and f_1.

In the following section, these steps are carried out in detail to prove the simplest case of the nonparametric version of Theorem 4.4. Here we complete the proof of Whitney's theorem by a more direct (if you like: averaged) adjustment of the $\tilde{f}_\tau(t)$, cf. [29, p. 256 et seq.]. Set

$$f_\tau(t) = \int_0^t h_\tau(s)\, ds - \frac{t}{2\pi} \int_0^{2\pi} h_\tau(s)\, ds.$$

This does indeed define a homotopy of C^1–maps between f_0 and f_1, since

$$h_\tau = f'_\tau \text{ and } \int_0^{2\pi} h_\tau(s)\, ds = 0 \text{ for } \tau = 0, 1.$$

We only need to verify that it is a regular homotopy, that is, $f'_\tau(t) \neq 0$ for all $\tau \in [0, 1]$ and $t \in [0, 2\pi]$. We have

$$f'_\tau(t) = h_\tau(t) - \frac{1}{2\pi} \int_0^{2\pi} h_\tau(s)\, ds.$$

If $h_\tau(t)$, for fixed τ, is not a constant in t, then

$$\left\| \frac{1}{2\pi} \int_0^{2\pi} h_\tau(s)\, ds \right\| < 1 = \|h_\tau(t)\|,$$

where $\|.\|$ denotes the euclidean norm in $\mathbb{R}^2$. So under this condition on $h_\tau(t)$ we are done; this holds in particular if $n_0 = n_1 \neq 0$. In the case $n_0 = n_1 = 0$ we need to show that the homotopy h_τ between the nonconstant maps $h_0, h_1 \colon S^1 \to S^1$ of degree zero can be chosen as a family of nonconstant maps.

Let $h_0^*, h_1^* \colon S^1 \to \mathbb{R}$ be lifts of h_0, h_1 (with respect to the standard covering $\mathbb{R} \to S^1$, $t \mapsto e^{it}$); these lifts exist for $n_0 = n_1 = 0$. By a regular homotopy of f_0 and f_1 we may assume that $h_0^*(t) = h_1^*(t)$ for

$t \in [0, \varepsilon]$ and some small $\varepsilon > 0$, and that h_0^* and h_1^* are nonconstant there. Then

$$h_\tau(t) = \exp(i((1-\tau)h_0^*(t) + \tau h_1^*(t)))$$

defines a homotopy with the desired properties. (In fact, it would be sufficient to require $h_0^*(0) = h_1^*(0)$ and $\max h_0^*(t) > 0$, $\max h_1^*(t) > 0$.)

4.2. Proof of the simplest case

We now illustrate the method of convex integration by proving the simplest case of Theorem 4.4, following [23]. We consider $(M, W) = (I, \partial I)$, where I is the interval $[0,1]$, and the trivial bundle $E = I \times \mathbb{R}^q$. Moreover, we assume that our differential relation is a product $\mathcal{R} = E \times A \subset E \times \mathbb{R}^q = E^1$. Without loss of generality we may assume that A is connnected. The conditions in Gromov's theorem then stipulate that A be open and ample, i.e. $\mathrm{Conv}(A) = \mathbb{R}^q$. We shall only prove the non-parametric h-principle; an additional point not contained in our general statement will be that we indicate how to obtain a solution C^0-close to the original non-holonomic section. Notice that this case covers the illustrative example of curves in $\mathbb{R}^3$ described in the preceding section.

Thus, let σ_0 be a solution of $\mathcal{R}$ defined near ∂I, and let φ be an element of $\Gamma(\mathcal{R}, \sigma_0, \partial I)$. We write this as $\varphi(t) = (\varphi_0(t), \psi_0(t))$ with

$$\varphi_0 \in C^0(I, \mathbb{R}^q), \quad \psi_0 \in C^0(I, A),$$

and this is supposed to satisfy

$$(\varphi_0(t), \psi_0(t)) = (\sigma_0(t), \sigma_0'(t)) \text{ near } t = 0, 1.$$

Our aim is to find $\sigma_1 \in \Gamma_0(E, \sigma_0, \partial I)$ with $\sigma_1 \simeq \varphi_0$ relative to a neighbourhood of $\partial I \subset I$ (which is automatic, since σ_1 and φ_0 are paths in $\mathbb{R}^q$ coinciding near the endpoints) and $\sigma_1' \simeq \psi_0$ as paths in A, again relative to a neighbourhood of $\partial I \subset I$. Moreover, we want σ_1 to be C^0-close to φ_0.

Since A is ample, we can find a step function $\chi: I \to A$ such that

$$\sigma_0(1) - \sigma_0(0) = \int_0^1 \chi(s)\, ds.$$

The fact that A is open and connected allows us to approximate χ by a continuous, contractible loop $\psi_1 \colon (I, \partial I) \to (A, \sigma_0'(1))$ with

$$\psi_1(t) = \sigma_0'(t) \text{ near } t = 1$$

and

$$\int_0^1 \psi_1(s)\,ds \text{ arbitrarily close to } \sigma_0(1) - \sigma_0(0).$$

Define $\widetilde{\psi} \colon I \to A$ as ψ_0 followed by ψ_1, parametrised in such a way that

$$\int_0^1 \widetilde{\psi}(s)\,ds \text{ is arbitrarily close to } \sigma_0(1) - \sigma_0(0),$$

and with

$$\widetilde{\psi}(t) = \sigma_0'(t) \text{ near } t = 0, 1.$$

Since A is open, this $\widetilde{\psi}$ may be adjusted, e.g. on a subinterval where $\widetilde{\psi}$ is constant and equal to χ, to $\psi \colon I \to A$ with

$$\int_0^1 \psi(s)\,ds = \sigma_0(1) - \sigma_0(0)$$

and still

$$\psi(t) = \sigma_0'(t) \text{ near } t = 0, 1.$$

Now define

$$\sigma_1(t) = \sigma_0(0) + \int_0^t \psi(s)\,ds \in \Gamma_0(E, \sigma_0, \partial I).$$

Since ψ_1 was chosen as a contractible loop in A and equal to $\sigma_0'(t) = \psi_0(t)$ for t near 1, we have $\sigma_1' = \psi \simeq \psi_0$ as paths in A, relative to a neighbourhood of $\partial I \subset I$. Thus, σ_1 has all the desired properties, except that it may not be C^0–close to φ_0. In order to achieve that, instead of tagging on a large loop ψ_1 to the end of ψ_0, one adds several smaller loops at intermediate points $t_\nu \in I$, with integral close to $\sigma_0(t_\nu) - \sigma_0(t_{\nu-1})$.

This completes the proof of Theorem 4.4 for the case under consideration. □

As a first application, we once again prove the Smale-Hirsch theorem for immersions $M^m \to N^q$ in the case of extra dimension $m < q$. Thus we consider $E = M \times N$, so that $(p_0^1)^{-1}(x, y)$ can be thought of as the space of linear maps $T_x M \to T_y N$, which after introducing

4.3. APPLICATIONS TO SYMPLECTIC AND CONTACT GEOMETRY

local coordinates may be identified with the space $\mathcal{M}^{q\times m}$ of $(q \times m)$–matrices. Then the principal subspaces for the coordinate direction x^i are of the form

$$P = \{(a_{kj}) \in \mathcal{M}^{q\times m}: a_{kj} = b_{kj} \text{ fixed for } j \neq i\}.$$

Moreover,

$$\mathcal{I}mm \cap (p_0^1)^{-1}(x,y) = \{(a_{kj}) \in \mathcal{M}^{q\times m} \text{ of rank } m\}.$$

Hence, writing b_j for the column vector $(b_{1j}, \ldots, b_{qj})^t$,

$$\mathcal{I}mm \cap P = \begin{cases} \emptyset & \text{if the } b_j, 1 \leq j \leq m, j \neq i \text{ are linearly dependent,} \\ \mathbb{R}^q - \text{span}_{j \neq i}(b_j) = \mathbb{R}^q - \mathbb{R}^{m-1} & \text{otherwise.} \end{cases}$$

So the intersection of the immersion relation with any principal space is either empty or the complement of an affine subspace of codimension at least 2, and hence ample. The Smale-Hirsch theorem follows.

4.3. Applications to symplectic and contact geometry

One application of the h-principle for ample relations concerns the construction of symplectic and contact forms "in codimension 1". The following proposition deals with the simplest instance of this construction.

PROPOSITION 4.6. *Let M be a 3–dimensional manifold, ω a nondegenerate 2–form on M (which in this dimension simply means a nowhere vanishing form), and $a \in H^2(M;\mathbb{R})$. Then ω is homotopic, through nondegenerate 2–forms, to a closed, nondegenerate 2-form ω' with $[\omega'] = a$.*

PROOF. Set $E = T^*M$ and consider the commutative diagram

$$\begin{array}{ccc} E^1 & \xrightarrow{\Delta} & \Lambda^2 T^*M \\ {\scriptstyle p^1}\downarrow & & \downarrow \\ M & \xrightarrow{\text{id}} & M \end{array}$$

from Lemma 2.4. Let λ be a closed 2–form on M with $[\lambda] = a$. Define a differential relation $\mathcal{R} \subset E^1$ by setting

$$\mathcal{R} = \{u \in E^1 : \lambda + \Delta u \text{ is nondegenerate}\}.$$

Let x^1, x^2, x^3 be local coordinates on M, and write a 1–jet in these coordinates as

$$j_{x_0}^1 \alpha = (dx^1, dx^2, dx^3) \begin{pmatrix} a_1 \\ a_2 \\ a_3 \end{pmatrix} + (dx^1, dx^2, dx^3) A \begin{pmatrix} x^1 - x_0^1 \\ x^2 - x_0^2 \\ x^3 - x_0^3 \end{pmatrix},$$

Then

$$\begin{aligned} \Delta j_{x_0}^1 \alpha &= (dx^1, \ldots, dx^m) A \begin{pmatrix} dx^1 \\ \vdots \\ dx^m \end{pmatrix} \\ &= \sum_{i<k} (a_{ik} - a_{ki}) \, dx^i \wedge dx^k. \end{aligned}$$

We also write λ in these local coordinates as

$$\lambda_{x_0} = \sum_{i<k} \lambda_{ik} dx^i \wedge dx^k.$$

Hence

$$\begin{aligned} \lambda_x + \Delta j_x^1 \alpha &= (\lambda_{12} + a_{12} - a_{21}) \, dx^1 \wedge dx^2 \\ &\quad + (\lambda_{13} + a_{13} - a_{31}) \, dx^1 \wedge dx^3 \\ &\quad + (\lambda_{23} + a_{23} - a_{32}) \, dx^2 \wedge dx^3. \end{aligned}$$

Now let P be a principal subspace in E^1 in the first coordinate direction, say. Such a principal subspace is given by setting a_{ik} equal to certain constants for $k = 2, 3$, in other words, the free variables in P are a_{11}, a_{21}, a_{31}. So as affine variables in P we may take

$$p_1 = a_{11}, \quad p_2 = \lambda_{12} + a_{12} - a_{21}, \quad p_3 = \lambda_{13} + a_{13} - a_{31}.$$

It follows that

$$\mathcal{R} \cap P = \begin{cases} P & \text{if } \lambda_{23} + a_{23} - a_{31} \neq 0, \\ \{b_1^2 + b_2^2 \neq 0\} & \text{otherwise.} \end{cases}$$

Thus, we either get the full space P, or otherwise the complement of the line $\{b_1 = b_2 = 0\}$ in the 3–dimensional affine space P. This shows that $\mathcal{R} \cap P$ is ample, so $\mathcal{R}$ satisfies the h-principle.

Now consider the 2–form $\omega - \lambda$ as a continuous (in fact, smooth) section of the bundle $\Lambda^2 T^* M \to M$. Since $\lambda + (\omega - \lambda) = \omega$ is nondegenerate and the fibres of Δ are contractible by Lemma 2.4, the 2–form

$\omega - \lambda$ lifts to a section $\varphi_0 \in \Gamma \mathcal{R}$. By Gromov's Theorem 4.4, this φ_0 is homotopic a holonomic section $\varphi_1 = j^1\alpha \in \Gamma \mathcal{R}$ (that is, with $\alpha \in \Gamma_0 E$) through sections $\varphi_t \in \Gamma \mathcal{R}$ for $t \in [0,1]$. Define $\omega_1 = \lambda + \Delta j^1 \alpha = \lambda + d\alpha$. This 2–form is nondegenerate, and a homotopy through nondegenerate 2–forms between $\omega_0 = \omega$ and ω_1 is given by $\omega_t = \lambda + \Delta \varphi_t$.

Notice that, *a priori*, we only have $\alpha \in \Gamma^1(T^*M)$. But this can be C^1–approximated by a smooth 1–form $\alpha' \in \Gamma^\infty(T^*M)$ in such a way that the linear homotopy between α and α' gives a homotopy from ω_1 to the smooth 2–form $\omega' = \lambda + d\alpha'$ through nondegenerate 2–forms. The 2–form ω' is clearly closed and realises the cohomology class $[\omega'] = [\lambda] = a$. $\square$

By similar arguments one proves the following generalisations of this result, see [**23**], where we write $\Omega^k(M) = \Gamma^\infty(\Lambda^k T^*M)$ for the space of differential k–forms.

(i) Given M with $\dim M = 2n+1$ and $\omega \in \Omega^2(M)$ with $\omega^n \neq 0$, there is an $\omega' \in \Omega^2(M)$ such that $d\omega' = 0$ and $(\omega')^n \neq 0$. Again, one may prescribe the cohomology class $[\omega']$.

(ii) Given M with $\dim M = 2n$, as well as differential forms $\alpha \in \Omega^1(M)$ and $\beta \in \Omega^2(M)$ with $\alpha \wedge \beta^{n-1} \neq 0$, there exists an $\alpha' \in \Omega^1(M)$ such that $\alpha' \wedge (d\alpha')^{n-1} \neq 0$.

There are also relative versions of these results available. This has consequences for certain equivalence relations between symplectic (resp. contact) forms, see [**23**] and also [**13**].

Closely related to these results is the construction of divergence free vector fields. Recall that if μ is a given volume form on a manifold M, then a vector field $X \in \Gamma^\infty(TM)$ is **divergence free** if the Lie derivative of μ with respect to X vanishes. By the Cartan formula for the Lie derivative, $L_X = d \circ i_X + i_X \circ d$, this is equivalent to $d(i_X \mu) = 0$.

Thus, k–tuples of pointwise linearly independent divergence free vector fields $X_1, \ldots, X_k$ are in one-to-one correspondence with k–tuples of pointwise linearly independent closed $(n-1)$–forms (where $n = \dim M$). Such forms can be constructed via the convex integration method from pointwise linearly independent, but not necessarily closed $(n-1)$–forms, see [**15**, p. 182], [**28**, p. 64].

I conclude these notes with a brief discussion of a result due to J. Gonzalo and myself [**12**] that brings us back to contact geometry proper, and that at first sight looks very similar to the problem of constructing divergence free vector fields. Recall that the *Reeb vector field* R_α of a contact form α is defined by the equations $d\alpha(R_\alpha, .) \equiv 0$ and $\alpha(R_\alpha) \equiv 1$. It is well-known that every closed, orientable 3–manifold is parallelisable; the following theorem says that one has a parallelisation in the contact geometric setting.

THEOREM 4.7. *Let M be a closed, orientable 3–manifold. Then M admits a triple of pointwise linearly independent contact forms $\alpha_1, \alpha_2, \alpha_3$ with pointwise linearly independent Reeb vector fields R_1, R_2, R_3.*

SKETCH PROOF. Observe that the requirement that the vector fields R_1, R_2, R_3 be pointwise linearly independent is equivalent to saying the same about the 2–forms $d\alpha_1, d\alpha_2, d\alpha_3$. Choose a contact form α_1 such that $\ker \alpha_1$ is a trivial 2–plane bundle. Such a contact form exists by the work of Lutz-Martinet and Eliashberg; for a direct proof see [**14**]. This amounts to the existence of a pair of 1–forms $\beta_0, \gamma_0 \in \Omega^1(M)$ satisfying the condition $\alpha_1 \wedge \beta_0 \wedge \gamma_0 \neq 0$.

Using a topological structure result for M and the h–principle for extensions, one shows that the problem can be reduced to a local extension problem. That is, one may work over local coordinate patches where α_1 is in Darboux normal form $\alpha_1 = dw - v\, du$.

We work with the bundle $E = T^*M \oplus T^*M$ and the relation $\mathcal{R}$ defined by

$$\mathcal{R}_x = \{(j_x^1\beta, j_x^1\gamma) \in E_x^1 \colon (d\alpha_1)_x, \Delta j_x^1\beta \text{ and } \Delta j_x^1\gamma$$

$$\text{are lin. independent and define a given orientation}\}.$$

With $x = (u_x, v_x, w_x)$, write the 1–jets in local coordinates as

$$j_x^1\beta = (du, dv, dw)\begin{pmatrix} b_1 \\ b_2 \\ b_3 \end{pmatrix} + (du, dv, dw)B\begin{pmatrix} u - u_x \\ v - v_x \\ w - w_x \end{pmatrix},$$

$$j_x^1\gamma = (du, dv, dw)\begin{pmatrix} c_1 \\ c_2 \\ c_3 \end{pmatrix} + (du, dv, dw)C\begin{pmatrix} u - u_x \\ v - v_x \\ w - w_x \end{pmatrix}.$$

4.3. APPLICATIONS TO SYMPLECTIC AND CONTACT GEOMETRY

Then

$$\Delta j_x^1 \beta = (du, dv, dw) B \begin{pmatrix} du \\ dv \\ dw \end{pmatrix},$$

$$\Delta j_x^1 \gamma = (du, dv, dw) C \begin{pmatrix} du \\ dv \\ dw \end{pmatrix}.$$

Employing a variant of Theorem 4.4, one only needs to prove the ampleness of the differential relation $\mathcal{R}$ in w–direction (this is essentially because we solve the problem step by step over small coordinate neighbourhoods). Notice that the principal subspaces in w–direction are 6–dimensional, with affine coordinates b_{i3}, c_{i3}, $i = 1, 2, 3$.

Assume that the orientation of $\Lambda^2 T^* M$ is given by the ordered triple

$$(du \wedge dv, du \wedge dw, dv \wedge dw).$$

Then, with $d\alpha_1 = du \wedge dv$ and

$$\Delta j_x^1 \beta = (b_{13} - b_{31}) \, du \wedge dw + (b_{23} - b_{32}) \, dv \wedge dw + (\dots) \, du \wedge dv,$$
$$\Delta j_x^1 \gamma = (c_{13} - c_{31}) \, du \wedge dw + (c_{23} - c_{32}) \, dv \wedge dw + (\dots) \, du \wedge dv,$$

we conclude that

$$\mathcal{R}_x = \{(b_{13} - b_{31})(c_{23} - c_{32}) - (b_{23} - b_{32})(c_{13} - c_{31}) > 0\}.$$

In a fixed principal subspace P, introduce affine coordinates s_1, s_2, s_3, s_4 defined by the equations

$$s_1 + s_3 = b_{13} - b_{31},$$
$$s_1 - s_3 = c_{23} - c_{32},$$
$$s_4 + s_2 = b_{23} - b_{32},$$
$$s_4 - s_2 = c_{13} - c_{31}.$$

Then the intersection of $\mathcal{R}$ with P takes the form

$$\mathcal{R} \cap P = \mathbb{R}^2_{b_{33}, c_{33}} \times \{s_1^2 + s_2^2 - s_3^2 - s_4^2 > 0\}.$$

Observe that $\mathcal{R} \cap P$ has the homotopy type of S^1. This information is used to show the existence of a (nonholonomic) section of $\mathcal{R}$. Moreover,

$\mathcal{R} \cap P$ is seen to be ample in P, so Theorem 4.4 (resp. the variant alluded to earlier) applies. We thus find 1–forms β, γ that are C^0–close to β_0, γ_0, respectively (so we can ensure $\alpha_1 \wedge \beta \wedge \gamma \neq 0$), and satisfy the condition that $d\alpha_1, d\beta, d\gamma$ be pointwise linearly independent.

Finally, we set $\alpha_2 = \alpha_1 + \varepsilon\beta$ and $\alpha_3 = \alpha_1 + \varepsilon\gamma$. For $\varepsilon > 0$ a sufficiently small real number, the 1–forms $\alpha_1, \alpha_2, \alpha_3$ provide a solution to our problem. □

Bibliography

[1] M. Adachi, *Embeddings and Immersions*, Transl. Math. Monogr. **124**, American Mathematical Society, Providence, 1993.

[2] G. Bredon, *Topology and Geometry*, Grad. Texts in Math. **139**, Springer, Berlin, 1993.

[3] M. Datta, Homotopy classification of strict contact immersions, *Ann. Global Anal. Geom.* **15** (1997), 211–219.

[4] M. Datta, A homotopy classification of symplectic immersions, in: *Symplectic Singularities and Geometry of Gauge Fields*, Banach Center Publ. **39**, Polish Acad. Sci., Warsaw (1997), 19–29.

[5] T. Duchamp, The classification of Legendre immersions, preprint (1982, revised 1996), available at http://www.math.washington.edu/~duchamp

[6] Y. Eliashberg, Classification of overtwisted contact structures on 3–manifolds, *Invent. Math.* **98** (1989), 623–637.

[7] Y. Eliashberg and N. M. Mishachev, Holonomic approximation and Gromov's h-principle, arXiv:math.SG/0101196.

[8] Y. Eliashberg and N. M. Mishachev, *Introduction to the h-Principle*, Grad. Stud. Math. **48**, American Mathematical Society, Providence, 2002.

[9] G. K. Francis and B. Morin, Arnold Shapiro's eversion of the sphere, *Math. Intelligencer* **2** (1979/80), 200–203.

[10] H. Geiges, Contact structures on 1–connected 5–manifolds, *Mathematika* **38** (1991), 303–311.

[11] H. Geiges, Applications of contact surgery, *Topology* **36** (1997), 1193–1220.

[12] H. Geiges and J. Gonzalo, An application of convex integration to contact geometry, *Trans. Amer. Math. Soc.* **348** (1996), 2139–2149.

[13] V. L. Ginzburg, Calculation of contact and symplectic cobordism groups, *Topology* **31** (1992), 767–773.

[14] J. Gonzalo, Branched covers and contact structures, *Proc. Amer. Math. Soc.* **101** (1987), 347–352.

[15] M. Gromov, *Partial Differential Relations*, Ergeb. Math. Grenzgeb. (3) **9**, Springer, Berlin, 1986.

[16] A. Haefliger, Lectures on the theorem of Gromov, in: *Proc. Liverpool Singularities Sympos. II*, Lecture Notes in Math. **209**, Springer, Berlin (1971), 128–141.

[17] M. W. Hirsch, *Differential Topology*, Grad. Texts in Math. **33**, Springer, Berlin, 1976.

[18] D. Husemoller, *Fibre Bundles*, Grad. Texts in Math. **20**, 3rd edition, Springer, Berlin, 1994.

[19] R. Kirby and L. Siebenmann, *Foundational Essays on Topological Manifolds, Smoothings and Triangulations*, Ann. of Math. Stud. **88**, Princeton University Press, 1977.

[20] A. A. Kosinski, *Differential Manifolds*, Pure Appl. Math. **138**, Academic Press, Boston, 1993.

[21] J. A. Lees, On the classification of Lagrange immersions, *Duke Math. J.* **43** (1976), 217–224.

[22] J. Lohkamp, Curvature h-principles, *Ann. of Math. (2)* **142** (1995) 457–498.

[23] D. McDuff, Applications of convex integration to symplectic and contact geometry, *Ann. Inst. Fourier (Grenoble)* **37** (1987), no. 1, 107–133.

[24] D. McDuff and D. Salamon, *Introduction to Symplectic Topology*, Oxford Math. Monogr., Oxford University Press, 1995.

[25] A. Phillips, Turning a sphere inside out, *Scientific American* **214** (1966), no. 5, 112–120.

[26] A. Phillips, Submersions of open manifolds, *Topology* **6** (1967), 171–206.

[27] D. J. Saunders, *The Geometry of Jet Bundles*, London Math. Soc. Lecture Note Ser. **142**, Cambridge University Press, 1989.

[28] D. Spring, *Convex Integration Theory. Solutions to the h-Principle in Geometry and Topology*, Monogr. Math. **92**, Birkhäuser, Basel, 1998.

[29] S. Sternberg, *Lectures on Differential Geometry*, Prentice-Hall, Englewood Cliffs, 1964.

[30] I. Ustilovsky, Infinitely many contact structures on S^{4m+1}, *Internat. Math. Res. Notices* (1999), 781–791.

[31] H. Whitney, Analytic extensions of differentiable functions defined in closed sets, *Trans. Amer. Math. Soc.* **36** (1934), 63–89.

[32] H. Whitney, On regular closed curves in the plane, *Compositio Math.* **4** (1937), 276–284.

Editorial Information

To be published in the *Memoirs*, a paper must be correct, new, nontrivial, and significant. Further, it must be well written and of interest to a substantial number of mathematicians. Piecemeal results, such as an inconclusive step toward an unproved major theorem or a minor variation on a known result, are in general not acceptable for publication. Papers appearing in *Memoirs* are generally longer than those appearing in *Transactions*, which shares the same editorial committee.

As of April 1, 2003, the backlog for this journal was approximately 4 volumes. This estimate is the result of dividing the number of manuscripts for this journal in the Providence office that have not yet gone to the printer on the above date by the average number of monographs per volume over the previous twelve months, reduced by the number of volumes published in four months (the time necessary for preparing a volume for the printer). (There are 6 volumes per year, each containing at least 4 numbers.)

A Consent to Publish and Copyright Agreement is required before a paper will be published in the *Memoirs*. After a paper is accepted for publication, the Providence office will send a Consent to Publish and Copyright Agreement to all authors of the paper. By submitting a paper to the *Memoirs*, authors certify that the results have not been submitted to nor are they under consideration for publication by another journal, conference proceedings, or similar publication.

Information for Authors

Memoirs are printed from camera copy fully prepared by the author. This means that the finished book will look exactly like the copy submitted.

The paper must contain a *descriptive title* and an *abstract* that summarizes the article in language suitable for workers in the general field (algebra, analysis, etc.). The *descriptive title* should be short, but informative; useless or vague phrases such as "some remarks about" or "concerning" should be avoided. The *abstract* should be at least one complete sentence, and at most 300 words. Included with the footnotes to the paper should be the 2000 *Mathematics Subject Classification* representing the primary and secondary subjects of the article. The classifications are accessible from www.ams.org/msc/. The list of classifications is also available in print starting with the 1999 annual index of *Mathematical Reviews*. The Mathematics Subject Classification footnote may be followed by a list of *key words and phrases* describing the subject matter of the article and taken from it. Journal abbreviations used in bibliographies are listed in the latest *Mathematical Reviews* annual index. The series abbreviations are also accessible from www.ams.org/publications/. To help in preparing and verifying references, the AMS offers MR Lookup, a Reference Tool for Linking, at www.ams.org/mrlookup/. When the manuscript is submitted, authors should supply the editor with electronic addresses if available. These will be printed after the postal address at the end of the article.

Electronically prepared manuscripts. The AMS encourages electronically prepared manuscripts, with a strong preference for $\mathcal{AMS}$-LaTeX. To this end, the Society has prepared $\mathcal{AMS}$-LaTeX author packages for each AMS publication. Author packages include instructions for preparing electronic manuscripts, the *AMS Author Handbook*, samples, and a style file that generates the particular design specifications of that publication series. Though $\mathcal{AMS}$-LaTeX is the highly preferred format of TeX, author packages are also available in $\mathcal{AMS}$-TeX.

Authors may retrieve an author package from e-MATH starting from `www.ams.org/tex/` or via FTP to `ftp.ams.org` (login as `anonymous`, enter username as password, and type `cd pub/author-info`). The *AMS Author Handbook* and the *Instruction Manual* are available in PDF format following the author packages link from `www.ams.org/tex/`. The author package can be obtained free of charge by sending email to `pub@ams.org` (Internet) or from the Publication Division, American Mathematical Society, 201 Charles St., Providence, RI 02904, USA. When requesting an author package, please specify $\mathcal{A}_{\mathcal{M}}\mathcal{S}$-LaTeX or $\mathcal{A}_{\mathcal{M}}\mathcal{S}$-TeX, Macintosh or IBM (3.5) format, and the publication in which your paper will appear. Please be sure to include your complete mailing address.

Sending electronic files. After acceptance, the source file(s) should be sent to the Providence office (this includes any TeX source file, any graphics files, and the DVI or PostScript file).

Before sending the source file, be sure you have proofread your paper carefully. The files you send must be the EXACT files used to generate the proof copy that was accepted for publication. For all publications, authors are required to send a printed copy of their paper, which exactly matches the copy approved for publication, along with any graphics that will appear in the paper.

TeX files may be submitted by email, FTP, or on diskette. The DVI file(s) and PostScript files should be submitted only by FTP or on diskette unless they are encoded properly to submit through email. (DVI files are binary and PostScript files tend to be very large.)

Electronically prepared manuscripts can be sent via email to `pub-submit@ams.org` (Internet). The subject line of the message should include the publication code to identify it as a Memoir. TeX source files, DVI files, and PostScript files can be transferred over the Internet by FTP to the Internet node `e-math.ams.org` (130.44.1.100).

Electronic graphics. Comprehensive instructions on preparing graphics are available at `www.ams.org/jourhtml/graphics.html`. A few of the major requirements are given here.

Submit files for graphics as EPS (Encapsulated PostScript) files. This includes graphics originated via a graphics application as well as scanned photographs or other computer-generated images. If this is not possible, TIFF files are acceptable as long as they can be opened in Adobe Photoshop or Illustrator. No matter what method was used to produce the graphic, it is necessary to provide a paper copy to the AMS.

Authors using graphics packages for the creation of electronic art should also avoid the use of any lines thinner than 0.5 points in width. Many graphics packages allow the user to specify a "hairline" for a very thin line. Hairlines often look acceptable when proofed on a typical laser printer. However, when produced on a high-resolution laser imagesetter, hairlines become nearly invisible and will be lost entirely in the final printing process.

Screens should be set to values between 15% and 85%. Screens which fall outside of this range are too light or too dark to print correctly. Variations of screens within a graphic should be no less than 10%.

Inquiries. Any inquiries concerning a paper that has been accepted for publication should be sent directly to the Electronic Prepress Department, American Mathematical Society, 201 Charles St., Providence, RI 02904, USA.

Editors

This journal is designed particularly for long research papers, normally at least 80 pages in length, and groups of cognate papers in pure and applied mathematics. Papers intended for publication in the *Memoirs* should be addressed to one of the following editors. In principle the Memoirs welcomes electronic submissions, and some of the editors, those whose names appear below with an asterisk (*), have indicated that they prefer them. However, editors reserve the right to request hard copies after papers have been submitted electronically. Authors are advised to make preliminary email inquiries to editors about whether they are likely to be able to handle submissions in a particular electronic form.

Algebra to KAREN E. SMITH, Department of Mathematics, University of Michigan, 525 University, Suite 2832, Ann Arbor, MI 48109-1109; email: `kesmith@lsa.umich.edu`

Algebraic geometry and commutative algebra to LAWRENCE EIN, Department of Mathematics, University of Illinois, 851 S. Morgan (M/C 249), Chicago, IL 60607-7045; email: `ein@uic.edu`

Algebraic topology and cohomology of groups to STEWART PRIDDY, Department of Mathematics, Northwestern University, 2033 Sheridan Road, Evanston, IL 60208-2730; email: `priddy@math.nwu.edu`

Combinatorics and Lie theory to SERGEY FOMIN, Department of Mathematics, University of Michigan, Ann Arbor, Michigan 48109-1109; email: `fomin@umich.edu`

Complex analysis and complex geometry to DUONG H. PHONG, Department of Mathematics, Columbia University, 2990 Broadway, New York, NY 10027-0029; email: `phong@math.columbia.edu`

*__Differential geometry and global analysis__ to LISA C. JEFFREY, Department of Mathematics, University of Toronto, 100 St. George St., Toronto, ON Canada M5S 3G3; email: `jeffrey@math.toronto.edu`

Dynamical systems and ergodic theory to ROBERT F. WILLIAMS, Department of Mathematics, University of Texas, Austin, Texas 78712-1082; email: `bob@math.utexas.edu`

Functional analysis and operator algebras to DAN VOICULESCU, Department of Mathematics, University of California, Berkeley, 970 Evans Hall, Floor 9, Berkeley, CA 94720-0001; email: `dvv@math.berkeley.edu`

Geometric topology, knot theory and hyperbolic geometry to ABIGAIL A. THOMPSON, Department of Mathematics, University of California, Davis, Davis, CA 95616-5224; email: `thompson@math.ucdavis.edu`

Harmonic analysis to ALEXANDER NAGEL, Department of Mathematics, University of Wisconsin, 480 Lincoln Drive, Madison, WI 53706-1313; email: `nagel@math.wisc.edu`

Harmonic analysis, representation theory, and Lie theory to ROBERT J. STANTON, Department of Mathematics, The Ohio State University, 231 West 18th Avenue, Columbus, OH 43210-1174; email: `stanton@math.ohio-state.edu`

*__Logic__ to THEODORE SLAMAN, Department of Mathematics, University of California, Berkeley, CA 94720-3840; email: `slaman@math.berkeley.edu`

Number theory to HAROLD G. DIAMOND, Department of Mathematics, University of Illinois, 1409 W. Green St., Urbana, IL 61801-2917; email: `diamond@math.uiuc.edu`

*__Ordinary differential equations, and applied mathematics__ to PETER W. BATES, Department of Mathematics, Michigan State University, East Lansing, MI 48824-1027; email: `peter@math.msu.edu`

*__Partial differential equations__ to PATRICIA E. BAUMAN, Department of Mathematics, Purdue University, West Lafayette, IN 47907-1395' email: `bauman@math.purdue.edu`

*__Probability and statistics__ to KRZYSZTOF BURDZY, Department of Mathematics, University of Washington, Box 354350, Seattle, Washington 98195-4350; email: `burdzy@math.washington.edu`

*__Real analysis and partial differential equations__ to DANIEL TATARU, Department of Mathematics, University of California, Berkeley, Berkeley, CA 94720; email: `tataru@math.berkeley.edu`

All other communications to the editors should be addressed to the Managing Editor, WILLIAM BECKNER, Department of Mathematics, University of Texas, Austin, TX 78712-1082; email: `beckner@math.utexas.edu`.

Titles in This Series

783 **Ethan Akin, Mike Hurley, and Judy A. Kennedy,** Dynamics of topologically generic homeomorphisms, 2003

782 **Masaaki Furusawa and Joseph A. Shalika,** On central critical values of the degree four L-functions for GSp(4): The Fundamental Lemma, 2003

781 **Marcin Bownik,** Anisotropic Hardy spaces and wavelets, 2003

780 **S. Marmi and D. Sauzin,** Quasianalytic monogenic solutions of a cohomological equation, 2003

779 **Hansjörg Geiges,** h-principles and flexibility in geometry, 2003

778 **David B. Massey,** Numerical control over complex analytic singularities, 2003

777 **Robert Lauter,** Pseudodifferential analysis on conformally compact spaces, 2003

776 **U. Haagerup, H. P. Rosenthal, and F. A. Sukochev,** Banach embedding properties of non-commutative L^p-spaces, 2003

775 **P. Lochak, J.-P. Marco, and D. Sauzin,** On the splitting of invariant manifolds in multidimensional near-integrable Hamiltonian systems, 2003

774 **Kai A. Behrend,** Derived ℓ-adic categories for algebraic stacks, 2003

773 **Robert M. Guralnick, Peter Müller, and Jan Saxl,** The rational function analogue of a question of Schur and exceptionality of permutation representations, 2003

772 **Katrina Barron,** The moduli space of $N=1$ superspheres with tubes and the sewing operation, 2003

771 **Shigenori Matsumoto,** Affine flows on 3-manifolds, 2003

770 **W. N. Everitt and L. Markus,** Elliptic partial differential operators and symplectic algebra, 2003

769 **Jie Wu,** Homotopy theory of the suspensions of the projective plane, 2003

768 **R. Höpfner and E. Löcherbach,** Limit theorems for null recurrent Markov processes, 2003

767 **Po Hu,** S-modules in the category of schemes, 2003

766 **Su Gao and Alexander S. Kechris,** On the classification of Polish metric spaces up to isometry, 2003

765 **Robert Bieri and Ross Geoghegan,** Connectivity properties of group actions on non-positively curved spaces, 2003

764 **J. Spandaw,** Noether-Lefschetz problems for degeneracy loci, 2003

763 **Yasuyuki Kachi and Eiichi Sato,** Segre's reflexivity and an inductive characterization os hyperquadrics, 2002

762 **Leiba Rodman, Ilya M. Spitkovsky, and Hugo Woerdeman,** Abstract band method via factorization, positive and band extensions of multivariable almost periodic matrix functions, and spectral estimation, 2002

761 **Oliver Druet and Emmanuel Hebey,** The AB program in geometric analysis : Sharp Sobolev inequalities and related problems, 2002

760 **Markus Banagl,** Extending intersection homology type invariants to non-Witt spaces, 2002

759 **Donald M. Davis,** From representation theory to homotopy groups, 2002

758 **Alan Forrest, John Hunton, and Johannes Kellendonk,** Topological invariants for projection method patterns, 2002

757 **Douglas Bowman,** q-difference operators, orthogonal polynomials, and symmetric expansions, 2002

756 **José Ignacio Cogolludo-Agustín,** Topological invariants of the complement to arrangements of rational plane curves, 2002

755 **M. A. Mandell and J. P. May,** Equivariant orthogonal spectra and S-modules, 2002

TITLES IN THIS SERIES

754 **Edward L. Green, Idun Reiten, and Øyvind Solberg,** Dualities on generalized Koszul algebras, 2002

753 **Daniel Panazzolo,** Desingularization of nilpotent singularities in families of planar vector fields, 2002

752 **Linus Kramer,** Homogeneous spaces, Tits buildings, and isoparametric hypersurfaces, 2002

751 **Bruce Allison, Georgia Benkart, and Yun Gao,** Lie algebras graded by the root systems BC_r, $r \geq 2$, 2002

750 **Masaki Izumi and Hideki Kosaki,** Kac algebras arising from composition of subfactors: General theory and classification, 2002

749 **Nanhua Xi,** The based ring of two-sided cells of affine Weyl groups of type $\widetilde{A}_{n-1}$, 2002

748 **Jürgen Ritter and Alfred Weiss,** The lifted root number conjecture and Iwasawa theory, 2002

747 **Armand Borel, Robert Friedman, and John W. Morgan,** Almost commuting elements in compact Lie groups, 2002

746 **Peter Niemann,** Some generalized Kac-Moody algebras with known root multiplicities, 2002

745 **Mikhail A. Lifshits and Werner Linde,** Approximation and entropy numbers of Volterra operators with application to Brownian motion, 2002

744 **Roger Chalkley,** Basic global relative invariants for homogeneous linear differential equations, 2002

743 **Heng Sun,** Spectral decomposition of a covering of $GL(r)$: the Borel case, 2002

742 **J. E. Gilbert, Y. S. Han, J. A. Hogan, J. D. Lakey, D. Weiland, and G. Weiss,** Smooth molecular functions and singular integral operators, 2002

741 **Francisco Santos,** Triangulations of oriented matroids, 2002

740 **Rick Durrett,** Mutual invadability implies coexistence in spatial models, 2002

739 **Georgios K. Alexopoulos,** Sub-Laplacians with drift on Lie groups of polynomial volume growth, 2002

738 **Yasuro Gon,** Generalized Whittaker functions on $SU(2,2)$ with respect to the Siegel parabolic subgroup, 2002

737 **Arjen Doelman, Robert A. Gardner, and Tasso J. Kaper,** A stability index analysis of 1-D patterns of the Gray-Scott model, 2002

736 **Wojciech Chachólski and Jérôme Scherer,** Homotopy theory of diagrams, 2002

735 **Martina Brück, Xi Du, Joonsang Park, and Chuu-Lian Terng,** The submanifold geometries associated to Grassmannian systems, 2002

734 **Michel Van den Bergh,** Blowing up of non-commutative smooth surfaces, 2001

733 **Milé Krajčevski,** Tilings of the plane, hyperbolic groups and small cancellation conditions, 2001

732 **Jan O. Kleppe, Juan C. Migliore, Rosa Miró-Roig, Uwe Nagel, and Chris Peterson,** Gorenstein liaison, complete intersection liaison invariants and unobstructedness, 2001

731 **Jesús Bastero, Mario Milman, and Francisco J. Ruiz,** On the connection between weighted norm inequalities, commutators and real interpolation, 2001

730 **Suhyoung Choi,** The decomposition and classification of radiant affine 3-manifolds, 2001

729 **Michael Grosser, Eva Farkas, Michael Kunzinger, and Roland Steinbauer,** On the foundations of nonlinear generalized functions I and II, 2001

For a complete list of titles in this series, visit the
AMS Bookstore at **www.ams.org/bookstore/**.

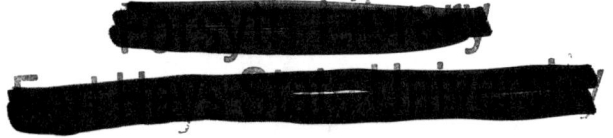